In Too Deep

In Too Deep

CLASS AND MOTHERING
IN A FLOODED COMMUNITY

Rachel Tolbert Kimbro

UNIVERSITY OF CALIFORNIA PRESS

University of California Press
Oakland, California

© 2022 Rachel Tolbert Kimbro

Library of Congress Cataloging-in-Publication Data

Names: Kimbro, Rachel Tolbert, 1978- author.
Title: In too deep : class and mothering in a flooded community / Rachel Tolbert Kimbro.
Description: Oakland, California : University of California Press, [2022] | Includes bibliographical references and index.
Identifiers: LCCN 2021033064 (print) | LCCN 2021033065 (ebook) | ISBN 9780520377721 (cloth) | ISBN 9780520377738 (paperback) | ISBN 9780520976436 (epub)
Subjects: LCSH: Neighborhoods—Texas—Houston—21st century. | Community life—Texas—Houston—21st century. | Mothers—Texas—Houston—Interviews. | Floods—Social aspects—Texas—Houston—21st century. | BISAC: SOCIAL SCIENCE / Social Classes & Economic Disparity | SCIENCE / Global Warming & Climate Change
Classification: LCC HN80.H8 K35 2022 (print) | LCC HN80.H8 (ebook) | DDC 307.3/362097641411—dc23
LC record available at https://lccn.loc.gov/2021033064
LC ebook record available at https://lccn.loc.gov/2021033065

Manufactured in the United States of America

31 30 29 28 27 26 25 24 23 22
10 9 8 7 6 5 4 3 2 1

To those who curated my own life:
Patricia, Bobbie, and Jean

Contents

	Acknowledgments	ix
	Introduction	1
1.	Choosing Bayou Oaks	16
	Are We in Pleasantville?	
2.	Storm Preparations	45
	I Had It All Planned Out before It Even Happened	
3.	During the Storm	78
	Get These Babies Out of the Water	
4.	Storm Recovery	109
	You Can Feel Sorry for Yourself When the Work's Done	
5.	Family Impacts	144
	This Past Year Has Really Been So Wretched	

6.	To Stay or Go	180
	Does Anyone Think This Is Crazy?	
	Conclusion	207
	Methodological Appendix	223
	Notes	233
	References	239
	Index	251

Acknowledgments

I am indebted to many people who helped see this book through to fruition, most especially the three dozen mothers and one elementary school principal who trusted me to tell their stories. They stopped their recovery labor for a few hours to talk to me, and did so again a year later. Their stories are unflinching, self-reflective, and full of grief, but also reflect some joy and plenty of dark humor. They are resilient and gracious. They show us what it is to be in community with others, and how women lead during difficult times. I pray they will not have to endure another flood.

Others read drafts of chapters, provided needed encouragement, and gave me critical but kind feedback, including Isabella Furth, Naomi Schneider, Sarah Damaske, Lori Peek, Alice Fothergill, Cayce Hughes, Anna Rhodes, Zhe Zhang, Elizabeth Long, Max Besbris, Elaine Howard Ecklund, Tony Brown, Caity Collins, Sarah Bowen, Dawn Dow, and the ladies of PEO Chapter E. The sociology departments at CU-Boulder and CU-Denver were the first scholarly audience for the work. Sandra Gray and Michelle Brown provided transcription and assured me that these stories were astonishing. Special thanks are due to Jim Elliott, who offered advice, helped me make connections, and listened to me endlessly, and to Sergio Chávez, who told me long ago I should have been a qualitative scholar and believed in me from the start. My graduate students—Laura Freeman Cenegy, Mackenzie Brewer, Catherine Boyd, Cassidy Castiglione, and Simon Fern—cheered me on and forgave me when writing took too much of my time. Rice University provided financial support for the project and also granted me a year's sabbatical for writing. And Elizabeth Keel, nanny, companion, and bodyguard extraordinaire, drove my children all over Houston to free up my time to write.

Finally, a word about my beloved family. My husband Rob and our teenagers Eleanor and Thomas put up with my absence during data collection and then with two years of frantic typing starting at 5:00 every morning. They listened to countless stories at dinnertime and endured my frustration and angst about the writing process. Their support and love, and their pride in my work, mean everything.

Introduction

Rebecca wasn't thinking about rain as she raced around her home on Memorial Day 2015, putting up decorations for her daughter's very pink first birthday party. In just a couple of hours, more than thirty guests would descend on her modest but stylish three-bedroom ranch in the Bayou Oaks neighborhood in southwest Houston. Her husband Paul was in his usual spot, the kitchen, making food for the party. Her 9-year-old son was staying out of the chaos in his backyard hideout, a playhouse equipped with a sleeping loft and real windows. Her daughter was busy crawling around underfoot and protesting that no one was paying attention to her during the preparations. During the party, some guests mentioned the approaching rainstorm, but mostly, between bites of pink-frosted cake and sips of champagne, they chattered about mutual friends and their elementary school community. This time of year in this affluent neighborhood, the social currency was advice about which teachers to request for next year. Although requesting specific teachers was technically not allowed by the local elementary school, everyone understood

1

that one should still try to get the best ones, and so the perceived attributes and deficits of each teacher were thoroughly discussed. Hours later, when the last of the guests had left and Rebecca had exchanged her daughter's pink party dress and party hat for pajamas, she and Paul decided to clean up in the morning. After all, it was late, and they were tired; the mess could wait.

As she lay in bed attempting to sleep, Rebecca wondered why some guests had been talking about the weather. It was just a rainstorm, right? But as the rain began to drive against the roof and swoosh rhythmically against the windows, the noise woke her up every hour or so. About 1:30 a.m., she made an annoyed Paul get up and check the house to make sure it was all right; it was. Rebecca got up herself at 3:00 a.m. to check, and everything was still fine. At 5:00 a.m. her phone rang. It was the school district saying that school was canceled. Bleary-eyed, she set her phone back on the nightstand and wondered why they were canceling school. Still confused, she sat up and swung her legs over the side of the bed—and felt the carpet squelch under her feet. "Paul! Get up! We have water in our house!" In disbelief, she and Paul slogged to the main part of the house. In the dining room, the plantation shutters on the windows were open a crack, letting moonlight in. Rebecca described what they saw:

> And so you could see—under the dining room table—the reflection of the table in the water. So it was three standing inches at that point. And as we're looking at it—and I'm kind of trying to wrap my head around this—one, like, solitary pink, Mylar balloon floats by in front of me. And I just looked at my husband and I went "womp-womp." It was like a scene out of, like, the cheesiest movie of all time. These pathetic little happy birthday streamers everywhere. And these little balloons in the water.

Over the next year, Rebecca and Paul faced trying times as they renovated their home with the help of their flood insurance. Although together the couple earned a middle-class income, it was still difficult to scrape together the cash reserves they needed for a temporary home, their continuing mortgage payments, and the advances to the contractors as they waited on insurance money. Then, in October, five months after the flood, Paul lost his job. Rebecca's income was not enough to sustain the family's expenses after the flood; as she put it, "He makes the living and I make the lifestyle." So they made careful use of the flood insurance payments to help them survive and get their home frugally finished. They moved back in seven months after the flood, right around Christmas 2015. Paul had trouble finding a new job and ended up being out of work for an entire year.

Throughout this time, Rebecca was determined to stay in Bayou Oaks. She'd selected this neighborhood carefully as the best place to raise her family. The homes were modest, and older, which was important to her as someone who valued style and architecture. The neighborhood was filled with young families. There was a park next to the local elementary school, which was only a few blocks from their home. And most importantly, the local schools, all the way through high school, were excellent. Selling and leaving never crossed her mind, even when the family experienced financial distress. But what would happen if they were flooded again?

Situated along one of the city's glorified drainage ditches that Houstonians call "bayous," the Bayou Oaks neighborhood in southwest Houston has never been a stranger to small-scale flooding. For residents of Bayou Oaks who lived within a block or so of the bayou, occasionally getting water in their sunken living room was just part of life. It was a trade-off they accepted for the chance to live in a coveted neighborhood, with a central location, great schools, and attractive mid-century modern homes. The Memorial Day flood of 2015, however, wasn't like those minor floods. The bayou's reach spread far south of the homes adjacent to its banks. Almost one-third of the Bayou Oaks homes zoned to Bayou Oaks Elementary flooded. The school itself took on a little water in an outlying building but was largely spared. Shaken but determined to persevere, the majority of flooded families in 2015 decided to renovate and stay in Bayou Oaks.

Memorial Day was a fluke, they reasoned. Unprecedented. Unlikely to happen again.

Then came the Tax Day flood of April 2016, which affected a much smaller number of homes but rattled the neighborhood more. What if this was going to keep happening? Some reflooded families had only just completed renovations from the first flood. A few decided they'd had enough. But most stayed.

The neighborhood rebuilt—again.

Then, sixteen months later, came Hurricane Harvey.

THE BOOK

This book tells the story of Bayou Oaks and its repetitive flooding, from the perspective of thirty-six mothers raising young children there. We follow the families across the course of more than a year, starting right after Hurricane Harvey flooded their homes and tracking them across the recovery year and beyond as they worked to restore their community for the third time in three years. Like other upper-middle-class and affluent mothers, this group of women prided themselves on the family lifestyles they managed and sustained. After Harvey, the mothers' roles as household logistics managers were strained to the breaking point as they labored to restore their homes and ensure their children's well-being. After Harvey, seventeen of the mothers, including Rebecca, were restoring their homes for the second or third time after flooding. Rebecca's story may seem like an outlier, but the world is full of Rebeccas, and more are in the making as climate change intensifies flooding events. How neighborhoods like Bayou Oaks respond to chronic flooding has major implications for the future of families and communities around the globe.

In Too Deep is a sociological exploration of what happens when climate change threatens the family life so carefully curated by upper-middle-class mothers. It argues that the careful choice of neighborhood and school is an overlooked component of the intensive mothering parenting style and demonstrates how mothers fare under the extreme stress of a year of disarray and dislocation, as well as the consequences of maintaining those high parenting standards for the mothers' mental and physical health and marital relationships. Instead of retreating after repeated flooding, the mothers dug in and sustained the community they had purposely chosen and nurtured, trying to keep social, emotional, and economic instability at bay. In the end, twenty-eight of the thirty-six mothers decided to stay in Bayou Oaks. As the "guardians of stability" for their families, the mothers ultimately argued that they were staying in the neighborhood because on balance it was the best environment for their children.

HURRICANE HARVEY

Hurricane Harvey made landfall as a category 4 about 10:00 p.m. on August 25, 2017, near Port Aransas, Texas. Its movement slowed to a crawl, and deadly rain pounded Texas—and especially Houston—for more than three days straight. Especially favorable meteorological conditions for torrential rain prevailed, and the unending downpour ended up dumping more than forty-three inches on Houston, about 85 percent of the average annual rainfall for the city. In some parts of Texas, more than sixty inches of rain fell. Nothing else in the recorded history of the United States comes close to that volume of rainfall, as Harvey was the wettest Atlantic hurricane ever recorded, and the National Weather Service had to update its map colors to show the unimaginable deluge.[1] More than half of the homes in Houston sustained damage from the hurricane, and 60 percent of those were not in any kind of floodplain.[2] Houston is no stranger to flooding. Its bayous snake through the city, mostly serving the purpose of draining the area during Houston's typical rainstorms. Houstonians are accustomed to staying off the roads during rainstorms, as the roads are designed to fill and drain into the bayous.

But the seventy-two-hour deluge during Harvey was like nothing ever witnessed, and the bayous and roadways were not built to withstand forty-plus inches of rain. For days, Houstonians were trapped in their houses—or if their houses were flooded, in neighbors' second stories. Local Houston meteorologist Jeff Linder estimated that 70 percent of Harris County, more than eighteen hundred square miles, was covered at one point by a foot and a half of water.[3] Ultimately, Hurricane Harvey's damage estimate of $125 billion ranked it second in US history in terms of cost, after Hurricane Katrina in 2005, according to the National Hurricane Center.[4]

Harvey was an equal-opportunity flood in Houston, at first. Higher- and lower-income neighborhoods were flooded across the city. Those neighborhoods' recoveries, of course, would not be equal opportunity. The higher-income neighborhoods inside floodplains, like Bayou Oaks, were populated by homeowners required to have flood insurance. The flooding's devastation was softened by the knowledge that eventually checks would be received to help make the homes whole again. The National Flood Insurance Program (NFIP), over time, has become an inefficient way to allow homeowners to purchase heavily subsidized insurance for their homes in floodplains. Although it was established with good intentions, to assist homeowners whose houses traditional insurance companies deemed too risky to insure, the NFIP now effectively encourages building and living in high-risk areas and relies on flood maps "that have only a vague connection to reality."[5] Higher-income neighborhoods in Houston *outside* of flood zones, where most of the homeowners lacked optional flood insurance, were in worse shape, but the homeowners tended to have more assets they could draw upon to rebuild or move. Lower-income neighborhood residents qualified for more immediate, short-term aid from the Federal Emergency Management Agency (FEMA), but renters were largely left out of larger and longer-term recovery resources.[6] In addition, lower-income Houstonians had more difficulty navigating the vast bureaucratic landscape post-Harvey. All of these factors help to explain why experiencing a disaster, even if it is felt equally across a community during a storm, may propel residents on vastly diverging trajectories in the days that follow.

Even in this city of flooded neighborhoods, after this unprecedented storm, Bayou Oaks stood out. Unprecedented? In Bayou Oaks, it felt like deja vu. While sea-level rises, perhaps a better-understood consequence of climate change, are projected to swamp coasts across the world, less well understood by the general public is how climate change also increases the intensity of rainfall events.[7] Flooding, now the most common disaster, is one major aspect of climate change that is on the rise in the United States and around the world.[8] Over time, especially with increased warming, "unprecedented" rainfall events are expected to occur with greater frequency, and floods are expected to become a reality for more and more parts of the world.[9] In all, more than forty million Americans are estimated to be at risk of catastrophic flooding.[10] But in Bayou Oaks, after two floods in two years, and now the threat of a third, the risk of home flooding from the hurricane was clear. Most of the residents undertook extensive preparations as Harvey approached, because they understood they were likely to be flooded again. Perhaps no other neighborhood in Houston had a clearer understanding of how a hurricane's flooding might inundate their homes. Those preparations likely saved some possessions and surely saved some lives, but in the end, the residents of Bayou Oaks had to confront yet another recovery process, with their resources depleted from dealing with the floods that had come before. Recovering from one flood is hard enough. But two or three? Bayou Oaks residents faced stark choices as Harvey's floodwaters started to recede, and this book explores why most of the mothers in the study decided to stay.

RESEARCH DESIGN

In Too Deep is based on seventy-two interviews with thirty-six mothers living in Bayou Oaks in 2017. As residents of a nearby, less expensive neighborhood, my children had previously attended Bayou Oaks Elementary as magnet students, so I had some familiarity with the neighborhood and both its strengths and troubles. I started by reaching out to the four flooded mothers I had stayed in contact with who still had children attending the school and asked them to refer me to others they believed I would not know, and who would not know each other. I first spoke to these mothers within the first eight weeks after Hurricane Harvey, and then again a year later in the fall of 2018, for a total of seventy-two interviews. This longitudinal approach allowed me to collect the mothers' flood stories while they were fresh, but also to follow up to see how the families fared after a year of regrouping. The Bayou Oaks mothers were an advantaged group. The median household income of the families in the study was $180,000, about three times the median household income in Houston. All but one mother had at least a college degree, all but two were married, and two-thirds worked outside the home, usually in professional occupations. While the families were affluent based on household income, their access to liquid assets had considerable variation, which would impact their recovery process and decisions about what to do with their homes. Thirty of the thirty-six mothers were White; the others were Black, Latina, and Asian. The mothers had between one and four children, with two children and two parents being the most common family size. In addition to participating in the two interviews spaced a year apart, many mothers wrote me emails or text messages in the intervening months to update me on their progress, and some sent me photos as well. I reached out to those I had not heard from. These efforts allowed me to keep tabs on each family's progress throughout the year. In addition to the interviews, I conducted participant observation at parties, community meetings, religious events, and school functions, which gave me a broader sense of the neighborhood and its social circles.

In the interviews, in addition to collecting the mothers' flood stories, I asked about their approaches to parenting, their relationships with their spouses and extended family, and their relationships with the local elementary school, and included detailed questions about why they had chosen Bayou Oaks and what they intended to do next. The interviews were semistructured, and there were some topics I made sure to cover; however, I allowed my interviewees to largely dictate the content and direction of the conversation. The stories are often terrifying; they are also heartwarming and funny. They are all gripping. Together, the two rounds of interviews provide an in-depth look at life after a disaster and identify the mechanisms working to hold the mothers to their community, the impact of the disaster on their families, and the strategies and schemas they drew on during the work of recovery and by which they measured success and failure. Additional methodological details are available in the methodological appendix.

CLASS AND DISASTER

Most of the disaster literature, for good reason, focuses on the experiences and responses of low-income communities. There is rich work showing how communities with fewer resources cope and recover (or don't). Lower-income disaster survivors don't just have a harder time with home recovery; their mental and physical health may also be differentially impacted by the disaster.[11] The recovery process for low-income people is also complicated by dislocation from local social networks, something that is not as disruptive for higher-income survivors. For example, after Katrina, poor families who relocated to Texas faced major challenges finding new homes and jobs after losing their local networks.[12] Additionally, higher-income and more-educated disaster survivors have an easier time navigating the bureaucratic procedures for postdisaster assistance.[13] So while there is a rich history of work documenting how less-advantaged communities respond to disaster, there is much less work examining the response of affluent communities. I argue that this is an important and overlooked area for research. While it might seem intuitive that disasters level social inequality within a community, disrupting everyone's lives equally, there is good community-level evidence that they actually *increase* social inequality.[14] Those who were well off before the disaster are even better off afterward. And those who were less well off before the disaster are worse off afterward. We can understand the latter—a disaster knocks us on our heels and requires resources to recover from. Jobs might be lost. Health problems and medical bills multiply. Evictions may occur. Higher-income people are more protected from these disaster insults. But why would more affluent communities be *better off* after a disaster? Unless we study affluent communities' responses to disaster, we cannot fully understand how a disaster serves to concentrate and amplify financial and social resources, leaving affluent communities better off. As we will see, the Bayou Oaks families received an outpouring of support, adding to their already substantial financial and social assets. This was on top of their substantial flood insurance payouts. Losing a home and most of one's possessions is devastating for any family. But the depth of resources affluent families have access to is understudied and often hidden from public view. Understanding these recovery dynamics may help better target resources after future disasters.

Further, although family and nonfamily households alike are impacted by disaster, studying households with children helps to illuminate inequity in disaster recovery. This is because income inequality in the United States is at its *worst* among families with children.[15] That is, income divides are greatest among American households when they include children. How families with children fare after disaster, particularly how the disaster itself and recovery from it impact parenting and family life, was a neglected area of research for decades.[16] A flurry of research after Hurricane Katrina in 2005 started to bring to light this overlooked component of the disaster experience. In particular, *Children of Katrina* highlighted the divergent recovery processes faced by families of different means, as well as the surprising autonomy and helpfulness of children in the recovery process.[17] In addition, the book, which followed families for seven years following the storm, revealed the importance of community institutions like schools in the recovery of families. Still, few studies have focused on the experience of parents after disasters.[18] Even within families, postdisaster trajectories may diverge, with adults and children on separate upward and downward emotional and resettlement trajectories.[19] Although we are beginning to understand how disasters impact children, and how disasters are experienced differently by families with children, there is still much to learn.[20] For the Bayou Oaks mothers, their high parenting standards were threatened by the flood and its aftermath, and their dedication to upholding those ideals caused considerable stress and anxiety across the recovery year.

INTENSIVE MOTHERING AND CURATED COMMUNITY

This book argues that the mothers' attachment to the neighborhood was so powerful because Bayou Oaks fit a series of precise criteria the mothers had for where to raise their families in this enormous, turbulent city, with the right kind of neighbors and school being primary. The mothers' choice of Bayou Oaks initially was a way to gain security for their families' lives in insecure times, specifically for their children. These mothers believed there was only one neighborhood in Houston's metro area of six million people that fit the bill: Bayou Oaks. And even after that security was repeatedly rocked by catastrophic flooding, almost all of the mothers *still* believed this to be the case. There was only one neighborhood for them, and they were committed to it. While parenting after a disaster has received more attention, particularly after Katrina, we know little about how mothers fit disaster parenting into their existing parenting schemas. Intensive mothering is a parenting style characterized by time-intensive, child-centered, emotionally involved, and expert-guided parenting.[21] Although this type of mothering is difficult to achieve, and is made easier by having more time and money, it nonetheless remains the dominant ideology by which mothers are evaluated.[22] This dominant ideology of mothering, which centers White, affluent mothering norms, is certainly complicated by differences in class and race.[23] Nonetheless, although this ideology may play out differently in practice, particularly for Black mothers, the emotionally intensive and child-centered approach to parenting may have more in common across class and race than is commonly believed.[24] That said, the specific components of intensive mothering differ, especially according to class. For instance, the "concerted cultivation" approach, which aims to produce accomplished young adults who will be advantaged in adulthood, is part of how White and affluent parents "opportunity hoard" on behalf of their children, necessarily shutting out other children.[25] Although usually not overtly malicious, this type of intensive parenting nonetheless works to maintain racial and class hierarchies.[26]

Across race and class, increasing economic instability pushes families to engage in "security projects," meaning actions and strategies that aim to situate their families favorably in an insecure world.[27] Surprisingly, even affluent families feel the pinch of economic insecurity, although they are best positioned to ride out economic turbulence. Sociologist Marianne Cooper describes this as living in different types of "risk climates," in which what one perceives as risky varies, but the feeling of threat does not.[28] Intensive mothering, then, can be thought of as an adaptive strategy in insecure times to mitigate the wide variety of risks that children face as they grow.[29] This "status safeguarding" is almost always undertaken by mothers, occurs across social classes, and is emotionally taxing.[30] In fact, when mothers take on these "security projects" with the goal of shoring up their children's futures, often in difficult circumstances, it produces a "motherload" of stress and anxiety.[31]

While in two-parent families both parents usually subscribe to the intensive parenting ethos, even when fathers sustain parts of its focused labor, they are usually shielded from its most time-consuming aspects.[32] With two-income families now the norm, this means that mothers are usually balancing paid work and caregiving work, while also trying to maintain intensive mothering ideals.[33] Despite the valorized image of mothers in our society and the immense concentration of time and labor mothers endure, motherhood remains a devalued status.[34] The combination of that devalued labor and the difficulty of balancing paid and caregiving work means something has to give. *In Too Deep* argues that some affluent parents reject some of the core tenets of intensive mothering while embracing others. Specifically, the mothers in Bayou Oaks chose that neighborhood as a way to short-circuit some of the demands of intensive mothering. As logistical heads of household and subscribers to the ethos of intensive mothering, they placed paramount importance on choosing the optimal setting in which to raise their children. By selecting Bayou Oaks and its elementary school intentionally, and by continuing to pour themselves into labor cultivating the community, the mothers actively curated the context in which their children would grow up. This maternal curation, a "security strategy," served to replace some of the intensive mothering labor the mothers would otherwise have had to engage in.[35] By selecting Bayou Oaks, the mothers were mitigating a variety of risks for their children. The safe streets, mostly White neighbors, good school with a racially diverse student body, and close-in location were all features the mothers valued and wanted to expose their children to. The mothers would spend time at Little League games, and they might hire a pitching or hitting coach on occasion. But they could rest easy knowing that their children were getting some of the best the city had to offer. The neighborhood itself provided family friends, religious institutions, great public schools, and art and cultural experiences.

What happens to intensive mothering, though, when disaster strikes? Do mothers maintain their extremely high standards, or do they drop them to focus on what really matters? Will they cling even more tightly to the neighborhood they have chosen so carefully? What if the very choice of the neighborhood, and particularly a decision to stay, now invites additional risks?

PREVIEW OF FINDINGS

The Bayou Oaks mothers poured themselves into disaster preparation and recovery just as intensively as they parented their children. They were eager to restabilize the neighborhood and their own family lives. Accustomed to managing work and a busy family life, the women dedicated themselves to getting their families back into their homes as quickly as possible. The typically gendered labor of family life became more stark during the recovery year.[36] The mothers reported being the person preparing the house before the storm, making decisions about evacuation and safety, managing contractors and laborers afterward, finding temporary housing, haggling with FEMA and insurance adjusters, going by the house daily to check on progress and get the mail, and being the point person to manage their children's emotional distress. Many saw this as a natural extension of their existing role: they were the family "logistics person" normally, so of course they would be during disaster recovery, too. But for most of the mothers, this overwhelmingly gendered recovery labor, on top of their regular work/family responsibilities, felt like "a second full-time job." As the mothers' mental load became unbearable, stress, health problems, and spousal conflict increased significantly. Several of the women contemplated divorce—sometimes because of financial stress, but more often because of their husbands' assumption that the wives would manage the recovery process just as they managed day-to-day family life. In other words, the couples had implicit agreements that the home was the wife's domain to maintain and sustain, even after a catastrophic event.

When their curated community is repeatedly threatened by catastrophic flooding, the mothers are not easily convinced to let it go. After all, the aspects of the neighborhood that drew them in the first place are still there. Again and again, the mothers told me that there were some practical reasons to stay, like finances, but that they still believed Bayou Oaks was the best place to raise their children. The mothers place a very high value on the neighborhood, its schools, and its community, and they do not want to lose it. They still live among neighbors who are "like them," their children are still zoned to the same schools, and they still have access to the city's amenities. Instead of retreating after repeated flooding for literally higher ground, the vast majority of the mothers dig in and work to sustain the community they have purposefully chosen and nurtured, trying to keep social, emotional, and economic instability at bay. As the guardians of stability for their families, the mothers ultimately argue that they are staying in the neighborhood because on balance it is the best environment for their children. In a chaotic, enormous city, they have found their ideal place to be—the place, they judge, that will give their children the best chance of future success—and it will take more than fifty inches of rain to change their minds.

1 Choosing Bayou Oaks

ARE WE IN PLEASANTVILLE?

On a sweltering "fall" afternoon in Houston in 2007, Nicole, age 38, was walking back to her car after working as an assistant in a class at a local community center. She and her husband, Dave, had recently sold their bungalow near downtown for three times what they'd paid for it a decade earlier. Now they were looking for a place with less traffic so their young son could play outside and eventually ride his bike. Nicole decided to drive through the neighborhood near the community center, Bayou Oaks, which she knew to be safe and in a great location not too far from the central city. As she drove through it, the place called to her. It looked like a real *neighborhood*, with wide sidewalks and mature oak trees, without being all the way out in Houston's far-flung suburbs. Homeowners were maintaining their lush lawns. The lots were enormous for the city, with plenty of space between neighbors. The homes were a mix of unassuming ranch houses and large mid-century modern homes. The big homes would be out of their price range, but because of the extraordinary appreciation of their starter home, they could now afford a ranch house in Bayou Oaks.

Nicole realized she was near the local elementary school, Bayou Oaks Elementary, and pulled over to watch as school let out. With a teaching certification, she had a special perspective on schools and now, of course, was starting to think about where their son would attend school. Without any prior knowledge of Bayou Oaks Elementary at all, she watched families pick up their children and liked what she saw. Many parents were walking their children home, instead of driving, and she saw "a good mix of skin colors," instead of the racially homogenous schools she knew were more common in the district. Bayou Oaks Elementary was a magnet school with a specialization in languages, and the odds of getting in via the Houston school district's lottery were only about one in four. As in many Houston Independent School District (HISD) magnet schools, however, residents of the school's attendance zone were essentially guaranteed admission. Nicole had no interest in playing the magnet lottery game; it felt like a gamble, and she thought it would be too stressful. So she had always intended to strategically buy into a good school's attendance zone before her son started kindergarten. Before she drove away, Nicole noticed that many parents were lingering after the bell rang, chatting with each other as their kids tore around in the grassy area in front of the school, darting in and out between large, hand-painted letters that spelled out B-A-Y-O-U O-A-K-S. This also seemed like a very good signal for the school—and for the neighborhood. Nicole wanted to be part of a community where parents hung out together and where the school was the center of the community.

As she drove home, Nicole knew she'd found their new neighborhood; she just had to let Dave know. It was exactly what they were looking for, with a great location, good school, and quiet, leafy streets where their son would one day ride his bike safely. It was the opposite of their current gritty, traffic-ridden neighborhood near downtown, but it was still close enough to the city amenities they valued, like museums and art galleries and restaurants. Ten years later, when I spoke with Nicole after Harvey had flooded their home, she still felt the same way. Her gut instinct that day, just driving around, had been right. Floods or no floods, Bayou Oaks was still the perfect neighborhood to raise a family. And they weren't going anywhere. Nicole's story illustrates how the Bayou Oaks mothers felt about their neighborhood. It had all the qualities they sought for a place to raise their children. Choosing Bayou Oaks, then, allowed the mothers to curate a lifestyle for their families that they desired. Living there would be advantageous for their children, and so living there was part of their identities as good mothers. What would it mean, then, if they were to leave?

Originally advertised in the *Houston Chronicle* in 1955 as a community where children could "go to the right schools, play with the right kind of companions," Bayou Oaks became a destination for a largely Jewish population who moved to the suburbs seeking bigger yards, bigger homes, and yes, distance from Houston's growing African American community, according to the Houston Jewish History Archive. Today, Bayou Oaks is somewhat more diverse, with a large concentration of Jewish and other non-Hispanic White families but also a small percentage of Hispanics, Asians, and African Americans. Together, non-Whites today make up about 15 percent of the neighborhood.[1] At the same time, Houston's extraordinary outward expansion has turned this once-suburban neighborhood into a "close-in," desirably located community, only about twenty minutes from downtown except during peak rush hour (a desirable trait given that most Houstonians live much farther away). Bayou Oaks was suburban by design, but now quasi-urban by virtue of its (relative) proximity to downtown.

Not only was the location convenient, and the home prices (from the mothers' perspective) relatively affordable, but Bayou Oaks fit their values of urban living, with a tight-knit, child-focused community, and its choice was an identity statement in a city where the majority of Whites chose the more distant suburbs. It allowed them to retain their image of themselves as city people who valued access to the city's amenities and rejected the suburbs' racial homogeneity, but their kids could still walk and bike to school on wide sidewalks under shady oaks. Most importantly, the neighborhood provided the kind of community they were looking for, which was hard to find in Houston's jumbled, sprawling urban landscape with no zoning. As a result, Bayou Oaks became so desirable that just a few years before Harvey hit, in order to purchase a home there, buyers had to swoop in before it actually went on the market, and this was true across price points. Before the first major flood in 2015, homes in the neighborhood ranged in cost from $300,000 for a real fixer-upper to more than a million dollars, with the median home price about $400,000. For comparison, in 2014, city-wide, the median home price was about $195,000. The wide range in Bayou Oaks prices before 2015 reflected the variation in housing stock, from modest three-bedroom ranches to five-bedroom, sprawling mid-century moderns, all built in the early 1950s.

The Bayou Oaks mothers were exemplars of the "concentrated cultivation" approach to middle-class parenting first described by Annette Lareau.[2] While Lareau's initial study focused mostly on the home environment, other researchers have explored how class influences parenting and family life in other arenas, notably the school environment.[3] I argue that an overlooked component of intensive mothering, especially for middle- and upper-middle-class families, is the choice of neighborhood.[4] While *concerted cultivation* refers to the child's personal development, *curation* refers to the child's environment, and I argue that mothers are actively curating their children's environment when they choose a neighborhood and a school. In most places, the neighborhood choice is also the school choice, making those decisions interlinked.[5] When parents choose a neighborhood, they are choosing a package of amenities for child-rearing. The ability to choose a neighborhood at all reflects class (and racial) privilege; the ability to do so with children's future success in mind reveals a way that inequality is reproduced in an era of increasing economic anxiety. The Bayou Oaks mothers wanted to situate their children in the best possible position socially, geographically, and educationally. While they believed parenting was important for children's development and exhibited some elements of intensive mothering, they also recognized the crucial nature of an environment where children could thrive and saw choosing that environment as part of their identities as "good mothers."

In Bayou Oaks, mothers expressed this sentiment over and over. It was seen as a given that they had to choose a place to live where the public schools were strong, preferably "all the way up," or through high school. That, they felt, left them with only a handful of neighborhood options unless they wanted to move to the suburbs. In addition to the schools, the mothers reported evaluating family lifestyle elements, such as commuting times, local amenities, and the presence of other families, in their choice of neighborhood. All of these factors, along with their financial resources, allowed the mothers to "curate" the type of family life they wanted for their children and for themselves. That curative labor was not complete when the family moved into Bayou Oaks, but continued as the mothers worked to build their social networks within the neighborhood, something made easier by the presence of many other families who were in a similar life stage.[6] Curation, then, was a constantly evolving, labor-intensive strategy the mothers used to situate their children for the best possible lifestyle. As we will see, once mothers believe their curation has resulted in a good place for their children to grow up, it will be difficult to change their minds.

The Bayou Oaks mothers frequently mentioned several features that had led them to choose this lifestyle and this neighborhood, all of which were framed as related to their parenting. First, they valued the close-in location of the neighborhood, rejecting the long commutes and homogeneity of the suburbs. Next, they lauded the people who lived in Bayou Oaks, distinguishing them both from suburbanites and from the wealthier residents in two nearby communities. Jewish mothers valued the proximity to major Jewish institutions. And most importantly, the mothers valued Bayou Oaks

Elementary as the place that would educate their children. In this chapter we learn why—and how carefully—the mothers chose where they wanted to live and raise their families. The choice of neighborhood was about the "parameters of their child's life," one that was freighted with symbolic weight.[7] The mothers believed that the qualities Bayou Oaks offered were unique in this fourth-largest city in the country, and their belief that the neighborhood was perfect for raising their children was barely shaken by the repeated flooding. Instead of departing en masse after the floods, they dug in and fought for the neighborhood's future and the lifestyle they had carefully curated within it for their families.

The mothers justified their choice of Bayou Oaks with several major narratives, including its location, the type of homes, who they did (and did not) want to be their neighbors, and its elementary school. This combination of factors, they felt, could not be found anywhere else in Houston. But more importantly, they believed that their choice of neighborhood was a statement about the type of people—and mothers—they were (and were not). The choice of Bayou Oaks, then, was not just tied up with their active curation of their children's lives, but also with their own identities.

The location of the neighborhood within Houston was mentioned by many of the mothers as a major reason they chose Bayou Oaks. Houston is an unusual city in many ways, but one distinctive feature is that the city's most exclusive neighborhoods are close to the central core of the city rather than in the suburbs. In addition, the lack of zoning and rapid growth of the city have led to urban sprawl across more than ten thousand square miles, an area larger than the state of New Jersey. People can drive an hour and a half and still be in "Houston." Suburban commutes on the city's notorious freeways can top two hours a day. While many families brave such commutes in exchange for the relative tranquility and large, much less expensive houses in the suburbs, some families seek to live as close to the center of the city as possible, both for easier commutes and for quicker access to the city's significant cultural amenities. Proximity to downtown and—crucially—the Texas Medical Center, the largest medical center in the world, which employs more than 100,000 Houstonians, makes Bayou Oaks the best-situated neighborhood in Houston where one can buy a home for under $800,000. It has the wide, flat streets characteristic of Texas; well-appointed landscaping; and majestic, mature live oaks. If you didn't know better when you drove the streets, you'd think you were in the suburbs. Of course, when the neighborhood was founded in the 1950s, it was the suburbs.

Proximity to the central city mattered to the mothers not just because it meant shorter commutes for themselves and their husbands, but also because it meant less time driving children to school and after-school activities—a common problem of American suburban life. Like others in their social class, this group of well-educated, affluent mothers factored in their children's extracurricular activities as just part of life, something that would be easier if they lived close to city amenities. While this was a practical concern—these were busy families, and there is only so much time in the day—it was also a value statement about how they wanted to spend their time, as Kelly, 48, put it:

> We chose it because of the proximity of the medical center. So we could get a lot more for our money farther out [in the suburbs], but we just didn't want to pull our family up. Being able to have our daughter go to school nearby, and being able to drop off in 10 minutes and not have to spend all our time in the car. Like we wanna spend time as a family and we just . . . we just love Bayou Oaks.

These mothers knew what they wanted from family life—and made sacrifices to achieve it. They wanted their children to have access to excellent dance teams and piano teachers, but not at the cost of excessive driving. They could be home from work in time to cook dinner or get their kids to baseball practice, and they could drive their kids to ballet without it taking forty-five minutes. This was not just about easing schedules; it was about enabling the kind of family life they valued. The extra time they gained, they envisioned, would be family time, as described by Laura, forty-two, a stay-at-home mother of two:

> And so with this choice and this location, it meant that [my husband] was gonna be home for dinner and actually spend a few hours with everyone. And you know, the trade-offs were a smaller home at a higher price point, but we felt like what we were getting for that, it was more than worth it.

The Bayou Oaks mothers were willing to live in a smaller, older, and more expensive home if it meant they could have the type of family life they desired, and they believed it was the best type of environment for child-rearing.

In addition to saving time through shorter commutes, in Bayou Oaks the mothers also would be well poised to take advantage of all the amenities of the city. It was important, they thought, for their children to have access to these institutions. They believed that if their families lived in the suburbs, they would not visit the city's restaurants and museums nearly as often, because it would take so long to get there. Michelle, forty-two, an interior designer and mother of two, appreciated all that Houston offered, and that living in Bayou Oaks allowed her easy access to it:

> I also love that it's affordable for what I consider in-city living. I think we're quite convenient to museums, and theater, and shopping, and the restaurants in Houston are bar none the best in the world. So I just . . . I'm a super huge cheerleader of Houston. But like, our neighborhood too.

Implicit in these statements about the choice of Bayou Oaks was a rejection of other types of neighborhoods and other types of neighbors. Many of the women believed that the neighborhood matched the type of people they were and the values they held. Like other house hunters, the Bayou Oaks families had relied on their within-class social networks for "signals" about which neighborhoods were good ones.[8] As Laura put it:

As [our realtor] got to know us better, he's like, "No you're gonna be happier closer into town with how you are with your family, with what you're looking to get out of the city, you're really gonna wanna be in Bayou Oaks." And we really value diversity and culture.

So their selection of the neighborhood was a value statement about themselves, but they also wanted to live around other people with the same values. And this meant definitely *not* living in the suburbs.

When I asked mothers why they had chosen Bayou Oaks as a place for their families, they almost always first offered up the neighborhoods they had rejected. While many found nearby, wealthier neighborhoods to be "snobby," they also did not find the idea of living outside of the city in its vast suburbs appealing. They considered themselves to be sophisticated urban women, and city living was a fundamental part of their own identities. Almost to a person, the mothers emphatically rejected the idea of living in the suburbs, many stating some version of the following, assuming I would know what they meant: "Well, we're definitely not suburbs people." The mothers felt that the suburbs weren't just a place to live; they were an embodiment of a particular type of person that they did not want to be. Instead, they were *city people* who embraced the whirl of urban life with all its diversity and action. This symbolic boundary between "city people" and "suburbs people" was, then, a critical part of the mothers' identities, and they put themselves squarely in the urban camp.[9] Dislike of the suburbs was so fundamental to Ellen, mother of two, that it was part of her cellular makeup: "Like, it's not in my DNA to live in the suburbs and commute."

Interestingly, several of the mothers indicated that while their husbands might prefer the suburbs, their preference had won out. The mothers were the guardians of stability for the household and the ones who researched neighborhoods and schools, so naturally they would decide where the family lived. *They* would decide what kind of environment was best suited for the raising of their children, and they would actively curate it. One mother also seemed to view suburban living as subscribing to a particular type of motherhood and wifehood that she did not endorse for herself: "The allure of a city isn't so alluring to [my husband], he'd be fine to move to [the suburbs]. I say to him, 'If you're gonna do that, just go find a new wife to do that with, 'cause I'm not going.'"

When I asked mothers to specify what they meant by saying they "were not suburbs people," they offered a few different narratives, framed primarily around who makes undesirable neighbors. Some felt strongly that people who choose to live in the suburbs are racist and live in the suburbs specifically because they don't want their children going to school with poor or minority children. Andrea, 49, explained: "I may be stereotyping, but I see people out in the suburbs as, you know, lots more white, and, and a little more racist, and a little more prejudiced, and a little more conservative." While Bayou Oaks was not as diverse as the mothers believed, just being "in the city" meant exposure to Houston's extraordinary ethnic diversity, and they liked the idea of living around other people who valued that diversity instead of running from it. Some said the suburbs were filled with conservative Republicans ("it's so red out there"), or people who want to "live in a bubble." For the mostly politically progressive mothers of Bayou Oaks, these were not appealing imaginary neighbors, perhaps especially in this period following the 2016 election season.

In addition to rejecting the suburbs for the people and the commutes, the mothers objected to the style of homes there. By and large, they believed Bayou Oaks's modest ranch homes and larger midcentury moderns were aesthetically pleasing and not "cookie-cutter," like the homes they pictured in the suburbs. There was widespread distaste for "gigantic McMansions," huge homes that all looked alike and had odd features like castle turrets and "ugly columns." Instead, Bayou Oaks mothers favored their smaller, more energy-efficient ranch-style homes, which retained "character" but were still meant for family living. Choosing Bayou Oaks, to them, was a rejection of excess, an embrace of a simpler life—one that valued time with family and friends over the size of one's kitchen. What went largely unacknowledged by the women I spoke with was the high cost of living near the city, especially in neighborhoods they would find desirable. While Bayou Oaks was one of the lower-cost neighborhoods near the center of the city, the median (pre-Harvey) value of the homes lived in by the mothers in my study was $450,000. To them, this was a reasonable amount to pay for a home with all the amenities they desired. It's important to note that in their worldview, when assessing two nearby communities with homes starting around $800,000, Bayou Oaks was a comparative bargain.

In addition to narratives about rejecting the suburbs because of neighbors, homes, and commutes, the mothers also provided narratives of wanting to expose their children to the world; in these narratives they often drew a contrast to their own upbringing. As Sarah, 42, put it:

> I've always kind of envisioned that kind of neighborhood where you can walk outside and talk to your neighbors. And kind of going back a little bit, it was really important to me having grown up in the suburbs here . . . having experienced the suburbs here, I'm like, "I'm not doing that to my kids." And that's something that my husband and I still clash about. He's like "Why are spending all of this money to live in this neighborhood, when we could get so much more house for so much less money." But it's important for me to live urban. We have access to a big city, which I think is important in terms of raising your children, but then you also have this sense of community that it is small. And you kind of get that suburb feel with your sidewalks, and people walking outside, and knowing your neighbors.

The Bayou Oaks mothers believed that the neighborhood offered an urban lifestyle, including proximity not just to cultural amenities but also to people of different backgrounds. They believed that children needed to be around people not like themselves in order to develop into good citizens and viewed their selection of the neighborhood as bound up in wanting this exposure for their kids. Many mothers offered some version of the following: "You have all types of people, right? So like we do have, like, Muslims. And we have Black people, and Hispanic people, and Asian people, and Indian people. And you know, I love that about our neighborhood."

While technically this statement is accurate, it is important to remember that Bayou Oaks is 85 percent non-Hispanic White, according to the US Census Bureau. As in other studies of mostly-White neighborhoods, the mothers consistently overestimated the diversity of their neighbors.[10] While the mothers may have had a rosier view of the diversity in Bayou Oaks than was warranted, they were sincere about the desire for their children to grow up outside of a suburban bubble. I also asked mothers why they didn't choose two other nearby neighborhoods that were similar demographically and also had good schools, but were considerably more expensive. I expected they would cite the high cost of homes in those neighborhoods, and some did. More, however, exclaimed that those upper-class social circles were just not their kind of people. Part of the curation of the environment the mothers wanted for their children involved finding others like themselves.

The mothers I interviewed believed that the people in Bayou Oaks were uniquely good neighbors, the type of people who would have you over for dinner and whom you'd trust to watch your kids on short notice. They accurately assessed the social class of Bayou Oaks as well off and highly educated, but as several mothers emphasized, their neighbors were not snobby. In fact, they symbolically distinguished Bayou Oaks residents not only from suburbanites, but also from the people in two nearby, wealthier areas: "I don't like it there . . . like, it's just not me. . . . I don't feel connected to that area, I feel like there's definitely a nose-up-in-the-air feeling that I just don't want to raise my children around. . . . I don't want that for my kids."

The mothers were thus focused on two types of exposure for their children:—a set of people they *did* want their children exposed to (a racially diverse urban population) and another set of people whom they did not want them exposed to (a racially homogenous, wealthy population, in both suburban and urban settings).[11] The choice of

Bayou Oaks, then, allowed them to calibrate those exposures according to their desires and to curate the social environment for their children. Mothers believed that Bayou Oaks, while full of educated professionals, also had a laid-back attitude relative to other, wealthier neighborhoods:

> I don't know what it is about this area, but it's ... it's my people. Like these are the people who ... who are like me. Like kinda like laid back ... you know. And so in this area is where I feel like we belong. I feel like [nearby neighborhood]'s kind of like snobby. Like I hear stories about [their public elementary school]. And that kind of intimidates me. And I don't ... I don't feel like I fit in with that crowd.

Allison, 48, a stay-at-home mother of two older children, appreciated that her neighbors weren't trying too hard:

> I liked that it was established without being really fussy. It wasn't so perfect and so kind of obviously material or ostentatious, with like that understated.... I felt like I could relate. [It was] just total projection that I would relate to the values of the area. You know, that people would care about good schools, and ... and family, and community. But it wasn't about being perfect.

In Bayou Oaks, they didn't have to be perfect. It was a place where these women felt comfortable, and where they wanted to raise their children, away from the monolithic suburbs but also away from the nearby neighborhoods which they felt were more into keeping up appearances. Bayou Oaks offered a balance between the carefully cultivated, high-intensity parenting ethos they ascribed to and a place where their kids could roam and play safely. This was a quality of the neighborhood lifestyle that was commonly mentioned, although it seemed as though that was more of an ideal than the reality. Prearranged playdates, as in other affluent urban and suburban families, were more common in the neighborhood than truly free-ranging children.[12] The mothers did feel safe, though, sending their kids alone by bike to friends' homes—something that has become uncommon in other upper-middle-class neighborhoods.[13] Tara, 34, mother of three, was looking for a place to raise her kids where they would have instant friends and got visible evidence of that when she visited Bayou Oaks: "We drove to Bayou Oaks and we saw all these kids out. And we thought this is where we wanna live. There's a street with kids, and [my daughter] can knock on the door and ask for someone to play with her." These were the qualities the mothers sought, and they thought that Bayou Oaks was a neighborhood for people like themselves—something they believed was hard to find: "Everybody's—you know—for the most part, kind of like we are." Mothers also valued and frequently mentioned the tight-knit nature of the community, which they believed was unusual in a big city. As Michelle put it:

> I love that it feels like we live in a small town within the fourth largest city in the country. You know when we go out to dinner, I run into people I know. I go to the grocery store, I run into people I know. And I love that. I did not have that growing up. And so it's kinda the best of both worlds.

In an era when the conventional wisdom is that meaningful interactions with neighbors are declining, it is notable that so many relationships and intraneighborhood social occasions were casually referred to by the mothers.[14] This may be because their neighbors were, in many cases, their friends, and Americans today are more likely to live around people that match their own characteristics and political leanings.[15] The mothers hung out with each other on a regular basis, formally and informally. Block parties were common, and the mothers described sitting out in their front yards on lawn chairs in the evenings, drinking wine with neighbors while they let their kids burn off energy. It seemed as though every weekend there was some neighborhood gathering: a cocktail party, a competitive gaming night at which teams wore matching T-shirts, a gala or auction for Bayou Oaks Elementary or one of the local sports clubs, or a book club meeting. One mother's annual "Christmakwanzakah" event, celebrating every major winter holiday, was always well attended. Although the heart of the social network in Bayou Oaks was formed by Bayou Oaks Elementary families, the network expanded out to those whose children were in private school as well as those whose children had long since graduated. Almost all the mothers in the study knew all of their immediate neighbors, and some knew who everyone was for several surrounding blocks. Local sports leagues and dance studios provided another avenue for parents to meet each other and spend time together outside the neighborhood. The local Little League, in particular, was well known for its tight-knit group of moms who spent long hours on the bleachers chatting.

While it is impossible to assess after the fact whether the social cohesion expressed by the mothers was a retroactive expression of having experienced three communal floods, the stories of neighborly interactions went back well before the first major flood. Sarah, who worked full time in public relations and had two young boys, reported a memorable introduction to the neighborhood after moving from a "rougher" area of Austin, Texas:

> I mean we were just so taken aback, 'cause we were living in Austin . . . centrally. And we didn't know some of our neighbors. And some of them parked on their grass, or probably were dealing drugs. You know? Fabulous things like that. And you know college kids were renting next door. And [when we moved here] it was just so foreign to us that people would be like coming up to us and being like, "Hi, I'm so-and-so, nice to meet you." And like, bringing us a bag of brownies. "Are we in Pleasantville? What is happening right now? Like why are these people being nice to us? You think these [brownies] are laced?"

Leah, an attorney and mother of two young children, recalled why they were initially drawn to Bayou Oaks in 2012:

> It's close enough to downtown but it still has a very nice community feel, you know, especially at the time that we were moving, there were tons of young families moving. The neighborhood had been a lot older for quite awhile, and so there was a lot of turnover, a lot of, you know,

rebuilding and remodeling of the older homes. And young families moving in with the intent of sending their kids to the school. It felt like it would be good to join that.

Clearly many of the mothers believed that the strong community found in Bayou Oaks existed before the first flood in 2015, and they could recount stories to illustrate that. It is also likely that the three floods, followed by the intense recovery and rebuilding process, served to both build and sustain the community feel in the neighborhood. Mothers believed they were "in this together" with their neighbors, and they deeply valued that sense of connection.

The web of social connections that stretched throughout Bayou Oaks was also significantly bolstered by Jewish residents and institutions. Bayou Oaks, as a historically Jewish community, is located near many of the city's major synagogues as well as the Jewish Community Center (JCC, or simply "The J"), a labyrinthine complex with scores of daily classes and activities, along with after-school programs and a day-care center. About one-third of the mothers in the study were Jewish, and for these mothers, it was a given that they would live in or near Bayou Oaks so they could be close to the Jewish institutions (and neighbors) they valued: "We like being near the city. We like being near downtown. And we like being near all of the Jewish institutions, which is our synagogue, the kosher stores, and the JCC, and most of the people that we know." In fact, most of the Jewish mothers told me that the Bayou Oaks Jewish community was the number one reason they had chosen the neighborhood in the first place, like Deborah: "We are Jewish. It has historically been a place where Jewish people live, although the Jewish community has dispersed some. And it was an affordable place to live with larger plots of land than other places with Jewish communities."

Some of the Jewish mothers had grown up in or near Bayou Oaks, but others were from neighborhoods and towns with few other Jewish residents and wanted something different for their own children. They didn't want their children to experience the same kind of social isolation they had felt growing up. Michelle was one of the mothers who'd grown up outside a Jewish community, and she was excited for her own kids to have Jewish friends and teachers:

> I grew up where there were no Jewish children. And I just never felt like.... I always felt like I was the only one. You know? And so when we moved to Houston, and we lived in this community that had all types of people and it was a very heavy Jewish community, I was like, gosh how nice is it that my kids can grow up here, like they might have a Jewish teacher. And they have Jewish kids in their class. And they ... you know it's just so different from the way I grew up. And I loved that I could provide them with that.

Even some of the non-Jewish mothers mentioned the Jewish community when explaining why they'd chosen Bayou Oaks:

> One of the core foundations of the Jewish faith is community. Like the requirement that you take care of each other and that's I mean "period." There's no ... there's no grey area about it. Having been such a nomad as a kid, that was important to me. And I was like okay, well that ... that's where I want to live, right? So that was a big one for me.

The sense of community experienced by Jews and non-Jews alike in Bayou Oaks was bolstered by active participation in the local Jewish institutions. Most residents without their own pool, regardless of faith tradition, joined the JCC's pool in the summertime and spent long, languid afternoons chatting while their kids splashed and played. Some participated in art or exercise classes offered at the center. Local grocery stores and bakeries catered to the Jewish population with kosher items. The fall Little League season, made possible by Houston's temperate weather, only had games on Sundays to avoid the Sabbath so that Orthodox Jewish children could participate. And the springtime bar and bat mitzvah scene was the 13-year-old's equivalent of the London season for Bayou Oaks kids, regardless of their faith tradition.

While perhaps not Pleasantville, within Houston's eclectic and transient mix of residents, Bayou Oaks did stand out. The mothers were not passive recipients of the community's benefits; rather, they spent a lot of time and energy on the care and feeding of that community. They recognized its importance for their children's healthy development and success, and they worked to curate it. They frequently spoke to their neighbors and primarily socialized within Bayou Oaks's boundaries. Constituting the core of the neighborhood's social infrastructure was Bayou Oaks Elementary, which had drawn many of the mothers to the neighborhood in the first place and in turn helped sustain the strong bonds that Bayou Oaks enjoyed.

Although the mothers cited many neighborhood attributes that had drawn them to Bayou Oaks, for most of the mothers the primary motivation for selecting Bayou Oaks was Bayou Oaks Elementary School. The parents viewed their decision about where to live as intricately linked to their decision about where their children should go to school, and they took great care in making that decision. Like other higher-income families, they viewed this decision as something that was permanent; they were looking for their "forever homes."[16] Houston has an unusual hybrid school choice system, in which children can attend their neighborhood school or can enter a lottery to attend a specialized magnet school. While intended to offer low-income students opportunities to attend higher-performing schools, in many urban areas school choice has brought the greatest benefits to more affluent students.[17] Because most magnet schools in Houston also have attendance zones, the result is a two-tiered system of admissions, in which families who can afford to own (or rent) a home in a particular attendance zone can essentially purchase a seat for their children at a particular school, bypassing the magnet lottery.

This system also results in many parents doing a good deal of research to ascertain the best school for their children—and therefore a disproportionate number of moves, particularly of middle-class and affluent families, occur close to the time that the first child starts school. These types of moves made explicitly to find good schools are especially pronounced in the United States among affluent families and when local income inequality is high—as it is in Houston.[18] Many of the Bayou Oaks families in the study moved into the neighborhood shortly before their first child started kindergarten or while they had toddlers. Only a few families moved in before they had any children, but all of these were intending to start a family, and most were already thinking about schools. Thus, children and their schooling were absolutely central in the decision-making about where to live, and the Bayou Oaks Elementary community—usually referred to as the "Bayou Oaks family"—formed the center of the mothers' social network.

For many of the families, the move to Bayou Oaks represented a move away from a central urban lifestyle. In Houston, unlike in some other large, urban school districts, a move to the suburbs is not necessary to access high-quality public schools, due to the considerable number of options.[19] So the mothers could move a bit farther from downtown but still maintain proximity to urban amenities and good schools. Mothers described moving from apartments and townhouses nearer downtown after they had children:

> When we first moved to Houston we lived [near downtown]. That was before children, and that was all fun, and exciting, and we walked places and then we had a baby, [in a townhouse] with three stories. And with a dog . . . and it just . . . you know it just wasn't conducive to like a family lifestyle.

Once they had made the decision to move to a "family" neighborhood, the Bayou Oaks mothers began doing research about elementary schools:

> So after law school, we did the whole young professional thing [near downtown]. Had one kid there. And then when number two was on the way, we started looking at, you know, elementary schools and where we wanted to raise our family. And there were maybe five or so elementary schools [that fit my criteria].

Like other upper-middle-class and affluent parents, the Bayou Oaks parents were adept at navigating bureaucracies to benefit their families, especially in the school choice environment.[20] They viewed the choice of where their children would attend school as something they could and should manage and control, and they weighed multiple factors in the course of that decision-making, seeking to maximize their children's outcomes.[21] The mothers took for granted that it was their responsibility to assemble and evaluate the available school options, and this was bound up with their conception of "good mothering," such that neighborhoods, schools, and parenting led to a "moral geography of mothering."[22] This moral geography, or the way that place intersects with the social construction of motherhood, meant that school reconnaissance, then, must be a key component of how the mothers decided where to live.

In addition to the necessary intel from their social networks, the reconnaissance phase of their neighborhood and school search focused on several components: a school's test scores and achievement; the feel of the place; and whether it matched the mothers' professed values of cosmopolitanism, public education, and diversity.[23] While most of the Bayou Oaks mothers reviewed test scores and school demographics online before choosing the neighborhood, they didn't stop there. Many mothers reported touring Bayou Oaks Elementary before deciding to buy a home in Bayou Oaks, because they wanted to get a feel for the place that couldn't be captured in official statistics or mission statements.

Prior to Harvey, the Bayou Oaks Elementary school building, built in 1960, was a low-slung, unassuming place. An elaborate mosaic depicting children of the world greeted families on the "porch," a covered patio at the front of the school. The interior gave a slight impression of a medical facility, with the walls tiled in institutional green and white and the unmistakable odor of crayons, sweaty children, and industrial cleaning chemicals. Flags of many nations hung in the lobby, along with analog clocks showing the time in Paris, Beijing, and Mexico City—representing the three cultures and languages the school offered instruction in. The bulletin boards that lined the hallways were filled with colorful children's art and writing. These cheerful decorations, along with the friendly office staff, lent warmth to the building that families quickly picked up on. Like Nicole, who had driven by the school when she was investigating the neighborhood, when Jennifer, 46, saw the school for the first time, she fell in love:

> When I went to visit Bayou Oaks Elementary and we toured around the school, I just got this magical feeling. I don't know how to describe it. It's just so. . . . it's so adorable. It's so warm and the parents are so involved. And the school is just. . . . I just fell in love with it. And I'm thinking, "All the parents in the school are so nice. They're my neighbors. I'm gonna like this neighborhood."

The idea that Bayou Oaks Elementary evoked a special feeling was repeated over and over by the mothers in my study; this was often referred to as "having drunk the Bayou Oaks Koolaid." As one mother put it:

> There's a feeling that you get when you walk into that place. Whether it's how the front greets you, or whether it's the fact that you don't feel like a stranger, or . . . you know? I don't even know what to tell you what it is. There is just an aura about that place.

A number of mothers reported not just taking a tour but also meeting with school staff before purchasing a home in the neighborhood. As Ruth, 36, mother of three, put it in her customary sardonic way, poking fun at herself:

> I sat down with Ashley [the principal]. You know I'm like a no-bullshit mother. I don't wanna meet with like, you know, the "entry people." I wanna meet with the head. I just do. I'm a Jewish woman. Like we have to . . . you know? And she sat down with me. There was not even a question.

Satisfied that the school and principal met her expectations, Ruth decided to enroll her children in the neighborhood school rather than sending them to private Jewish schools, as she and her husband had originally planned to do.

As they investigated the school, the mothers believed that Bayou Oaks Elementary matched their own personal identities as tolerant, open-minded city dwellers. By sending their children to this school, they were making a statement about what they believed was important for their children's first major educational environment, something they all took seriously. Most importantly, Bayou Oaks Elementary championed their values: by being a public school, by offering foreign-language instruction, and by its diverse student body. These qualities were important to the mothers both because they hoped to raise good humans and because they believed exposure to what the school offered would benefit their children's character in the long run. In addition, the choice of the school was a signal of their worthiness as moral mothers. In other words, this group of mothers did not just seek to maximize their children's academic success; they also hoped to maximize their moral success.[24] Like the extremely affluent parents in Rachel Sherman's *Uneasy Street*, the decidedly less affluent Bayou Oaks parents valued "exposure" to others for their children.[25] Unlike most of the Manhattan millionaire parents, however, the Bayou Oaks parents deliberately chose public schools to gain that exposure for their children.

These mothers largely rejected private schooling because public education fit with their identities as progressive urbanites. They valued public schooling so that their children would be exposed to a diverse group of children and because they believed in the concept of public schooling more generally. The mothers drew moral boundaries around what they chose (a racially diverse urban public school) and what the Whites in the suburbs chose (mostly White suburban schools). It was not only the White mothers who had these beliefs, however, although the reasons for seeking a racially diverse school environment were different.[26] Mary, 47, who was Latina, believed in public schooling and wanted to find a diverse school environment for her daughter, but felt limited by what she found in the mostly

White suburbs. Bayou Oaks and Bayou Oaks Elementary, though, fit the bill:

> You know we knew that we're both public school kids and we'd like to keep her in public school. We . . . know we would do private school if we need to, but we said we're gonna find an area that we know the public schools are strong and that way she would have the diversity of—you know—inner city.

Many of the families had the financial means to send their children to private schools but preferred a public school if they could find one that was acceptable: "I just wanted to live in a neighborhood that was close to the city without having to live in the city, and having access to a good school. I believe in the public school system. I think it's important to send your kids there if you can." This mother, like the others, believed that sending kids to public school was a collective responsibility and a statement about investing in the community. Of course, it was also a statement about her *own* identity as a good mother and good citizen.

Bayou Oaks Elementary, though, was not merely an acceptable public school option; it was an excellent school. It offered extensive language instruction, another feature the mothers highly valued. While it was not a language immersion school, children would receive exposure to Mandarin Chinese, Spanish, and French in kindergarten, then choose a language to focus on for the rest of their time in elementary school. The parents wanted a school experience that would cultivate their children to be citizens of the future world, which they believed would be diverse and where being multilingual would be an advantage. They valued diversity for its own sake and because they did not want their children to grow up "in a bubble," but also because they saw experience with diversity and languages as an advantage they wanted their children to have. As Sarah put it:

> Hearing different languages was important to me, especially living in Texas. I just wanted, you know, for my children to be exposed to a lot of people that didn't look like them, languages that they weren't

familiar with. So it was very intriguing. And then also I just had a gut feeling. . . . I had a vibe. You know?

Even more than the importance of public schools and the language curriculum, though, the mothers extolled the school's diversity. Unlike Bayou Oaks itself, which the mothers viewed as more racially diverse than it actually is, the elementary school is genuinely diverse, thanks in large part to its magnet program, which draws children from across the city. For example, in 2016 the school was 46 percent White, 15 percent Black, 24 percent Latino, 9 percent Asian, and 6 percent multiracial, and 24 percent of the students were classified as economically disadvantaged.[27] It is important to mention that this level of racial and ethnic heterogeneity is rare within the Houston school district, where only about 10 percent of all students enrolled are White, and most schools are majority Black or majority Latino. Given that other research has shown that White parents avoid urban schools with high minority enrollments and actively assess racial composition in school choice districts, it is perhaps not surprising that Bayou Oaks Elementary was the school of choice for these Houston parents, because it had what they deemed "the right mix" for its racial composition.[28] Like other affluent, well-educated families, the Bayou Oaks parents believed that an acceptably "diverse" school setting must be centered on a "large-enough" group of Whites or a "critical mass."[29]

It would be unfair, however, to assert that Bayou Oaks families were trying to avoid the racial and economic diversity in Houston's public schools like many other affluent Houstonians, who moved to the suburbs or sent their children to private schools. Indeed, the public schools in the two nearby, wealthier neighborhoods had much higher concentrations of White students and were not magnet schools. The Bayou Oaks mothers did not want to send their children to those schools because they were *too* White. Rather, their choice of Bayou Oaks Elementary—even with its relatively low levels of poverty and the high concentration of Whites within the district—was a distinctly countercultural signal.[30] Not only were these families determined to live in an urban area, but they also were settled on sending their children to public school in Houston—something the vast majority of well-off Houstonians avoided. Other sociologists have documented this phenomenon in other large cities; choosing a particular school, for these parents, is an important statement of their identities as "liberal urbanites" who defiantly choose to stay in the city and send their children to urban public schools, as long as the schools pass an acceptability threshold.[31] In Bayou Oaks, though, there was no sense of "risk" in choosing this urban school.[32] Instead, the mothers seemed delighted with the opportunity to send their children to Bayou Oaks Elementary.

The Bayou Oaks mothers saw diversity in the school population as a strength. They believed that if their children went to school with different types of people, the kids would both be able to function better as adults and also make the world a better place, because they would know how to get along with all types of people. I asked Nicole why it was important to her that her kids be exposed to children from other backgrounds:

> Because they need to know what the real world is like. [Bayou Oaks Elementary] is real world. And Houston to me is real world, even though it's still segregated. I'm not saying it's totally integrated, but I want them to grow up knowing that there are other people that are like them but don't look like them. And I'm just hoping that eventually we'll have—and it's slow—but eventually we'll have a generation that is so used to being around people that are totally look different from them, and have, you know, different family dynamics, and live in different kinds of living situations, and it's normal. You know? So that is something that's really important to me. I don't want them to just know about it over there. You know? I don't want them to say well yeah, there's people like that . . . they live over there. I want them to say yeah, I know people like that. Um so that, you know, just to make things more—I don't know—happier?

But the mothers believed it was important to precisely calibrate the amount of diversity their children experienced. The choice of

Bayou Oaks allowed the families to live among other affluent, mostly White neighbors but to send their children to a school with a significant proportion of low-income and minority children. Crucially, the ability to live in Bayou Oaks, and more broadly, to choose the type of neighborhood and school they most wanted, was enabled and reinforced by class and racial privilege. For this group of mothers, this neighborhood and this school were the best of both worlds, balancing during this "era of parental anxiety."[33] They believed that understanding different types of people, and being able to get along with them, was a critical advantage in life that they were providing their children. But there were limits. As one mother put it: "We want to expose them to reality. Just not TOO real."

The ability to navigate life "fluidly and with ease," of course, is a marker of race and class privilege.[34] Bayou Oaks Elementary, for this group of parents, offered the right amount of diversity, and evaluating the school's racial and ethnic composition was an important part of their decision-making criteria. Not all of the mothers sent their children to Bayou Oaks Elementary, instead choosing elite private schools and in two cases, different magnet schools. One of the handful of mothers who sent her children to private school was not impressed with the "virtue signaling" of the rest of the neighborhood: "These people in Bayou Oaks, they pat themselves on the back for sending their kids to public school. But it's really like they're paying for private school by buying into this neighborhood."

While this cynicism is perhaps warranted, the Bayou Oaks Elementary mothers were passionate about their school and their "school family." They thought carefully about the kind of school environment they wanted for their families, they rejected the idea of White flight, and most of them rejected the idea of private schooling (at least at the elementary level).[35] So while they largely believed that sending their children to public school was the right thing to do, the mothers also had found a golden needle in the Houston metroplex haystack: a school that met their precise ideas about the place that would help to inculcate their children with the right values. So they were able to send their children to exactly the kind of diverse, highly enriched school they wanted, while staying within the public system, which fit their priorities as social progressives. This phenomenon, of affluent, often White parents choosing urban schools for their children, certainly has pros and cons for school systems and for the non-White children who attend those schools.[36] If White parents see value in sending their children to racially diverse schools, this raises "difficult and uncomfortable" questions about who exactly is providing that value, and how participating uncritically in its valorization actually reinforces racial hierarchies.[37] For our purposes here, it is important to understand how the Bayou Oaks mothers thought about the school and neighborhood they'd found. They believed they were doing something socially responsible, something that would benefit their children, yes, but also ultimately the city as a whole. They were willing to invest their own time and money in a school community that they believed would benefit from that investment. Within their own social context, they were bucking the choices of other parents (save their neighborhood compatriots), and they were proud of those choices.

Once the mothers had settled on Bayou Oaks Elementary as their school of choice, they had to get their children admitted. As documented in other research about choosing schools and neighborhoods, these mothers learned quickly how to navigate the complex "school choice" system in Houston to best situate their children.[38] While they could take their chances and apply to excellent magnet schools through a lottery system, no matter where they lived, odds were low that their children would be accepted by the specific schools they wanted. If, however, they could manage to purchase or rent a home within the catchment area zoned to a particular school, they were likely to be able to enroll the children there. Enrollment was not guaranteed, as one mother who moved to Bayou Oaks shortly before kindergarten was horrified to discover. By that time in the summer, the entering classes were long full, but she managed to talk her way into a spot for her son in the English as a Second

Language (ESL) class thanks to a sympathetic principal. In general, as in this example, the local schools try to work with zoned families to get their kids in. Most of these mothers were unwilling to leave their children's education to a lottery. They targeted Bayou Oaks Elementary's attendance zone, paying hundreds of thousands more for homes inside the perimeter, so that their children could become part of the "Bayou Oaks Elementary family."

Purchasing a home in Bayou Oaks, then, gained these mothers access to the school they envisioned as perfect for their children, matching their own values as mothers but as citizens of a global city. Many of them paid top dollar to get it and avoided paying private school tuition or having to move to the suburbs. The high enrollment of local children in Bayou Oaks Elementary also meant that Bayou Oaks had another relatively rare feature in Houston: the children of families who lived there nearly all went to the same neighborhood school. In other Houston neighborhoods, children scatter far and wide on yellow school buses every morning. This truly neighborhood school helped to foster a sense of community among the neighbors, a sense they were seeking and valued. Just as in 1955, when the neighborhood debuted, the Bayou Oaks mothers hoped for the "right" schools and the "right" companions—although the definition of "right" had shifted over time to include diversity as a desirable feature. While, unlike in 1955, the Bayou Oaks mothers might not be fazed by a non-White neighbor, it is likely that their feeling that Bayou Oaks was for people "just like them" was at least partially bound up in its racial homogeneity.

Ultimately, most of the mothers felt that Bayou Oaks was as good as it got in Houston. If they rejected the suburbs and other, more exclusive neighborhoods, it was because nowhere else had the right combination of location, price, people, and schools. The choice of the neighborhood, for these mothers, was a shortcut to the kind of mothering they believed was important. By curating the right neighborhood and school environment for their children, they would give the kids access to the ideal set of opportunities and resources. Their children would be set up to thrive. Plus, by the time their children were enrolled in Bayou Oaks Elementary, the mothers were deeply invested, emotionally, socially, and financially, in Bayou Oaks. Once you've found this special combination of features, it's hard to let it go.

Even if it has one rather major recurring problem.

A year after Harvey, twice-flooded Rebecca reflected on her options and found them wanting:

> It's like where would I go? At the end of the day, where? Let's say I sold my house and I had cash in the bank, where would I go? There's no part of town that's gonna give me what I wanted in this neighborhood. That feels very neighborhoody, walking kids to school, where the streets aren't too busy, that I can afford, and that has good schools. There's not another part of Houston like this.

For these educated, urban, largely progressive mothers, Bayou Oaks was the ideal neighborhood in which to raise their children, and they were determined to fight to keep it that way.

2 Storm Preparations

I HAD IT ALL PLANNED OUT
BEFORE IT EVEN HAPPENED

I interviewed Julia, 40, for the first time a few weeks after Harvey in the townhouse she and her family were living in temporarily. Provided by a friend of a friend for an as-yet-unsettled but well below market value rent, the place felt more like a designer showroom than a family home. The house was just as its owner had left it: stuffed to the gills with valuable furniture and decorative items. I sat on a pristine white couch, balancing my coffee with some trepidation. Rich rugs were plush underfoot, and art in gilded frames hung on the walls. This traditional aesthetic was not to Julia's taste, she took pains to explain. In fact, it made her life harder; she constantly worried that her children would break or stain something in their temporary home. Dressed casually, sitting in a wingback chair with her legs tucked underneath her, Julia had just returned from chauffeuring her three children to school.

Over the course of the next three and a half hours, Julia's flood story unfolded rapid fire, punctuated by a healthy dose of black humor and an unusually perceptive sense of self. A part-time medical professional, Julia was a multitasker, and there was ample evidence of this during our interview. While we talked, we were briefly interrupted by a worker asking for her carpet choice for the renovation; another worker cleaning flood debris from Julia's dining room chairs, which were stacked in a tall pile against one wall; and several phone calls (which went unanswered). Also making an appearance was the family's housekeeper, who Julia explained was cleaning the townhouse for them even though the owner also had a cleaning service, because Julia didn't want her to lose income while the family was displaced.

Accustomed to managing logistics for her family, it was clear that Julia had attacked the postflood recovery process with vigor and skill. Her days were now filled with project management—managing her home's renovation, yes, but also keeping her family routine as normal as possible. It wasn't straightforward to get her kids to three different schools on time, in three different uniforms, with their appropriate homework, after losing everything in the flood and being displaced to an entirely different neighborhood. And that was just the morning routine. Similar to her approach to recovery, Julia's approach to Harvey preparations illustrates a typical pattern among the Bayou Oaks mothers: temporarily refocusing their intensive curation of family life on preventing flood damage and preparing their children emotionally for what was to come. In the days before Harvey hit, the mothers had adeptly prepared their homes, laid in supplies, and checked in with their neighbors, usually without much help from their husbands.

The weekend before Harvey, Julia was crafting Princess Leia headbands for her daughter's upcoming eighth birthday party. Over the summer, with her older two children at sleepaway camp, Julia and her youngest daughter Meg had planned the party to perfection. Dubbed a "Star Wars Field Day" party, it would feature Jedi robes, the aforementioned headbands, and Star Wars–themed games in the backyard with homemade prizes. Late in the weekend, Julia started hearing news reports about Harvey. With the party planned for the next Saturday, the day Harvey was projected to hit, she knew she had a problem. But Julia wasn't just worried about the party being impacted by the storm; she *knew* her home was going to be flooded. After the floods in 2015 and 2016, this is the reality that Bayou Oaks families live with when a big storm is predicted. Some Bayou Oaks mothers would have explained the situation to their daughter, dried some tears, postponed the party, and prepared for the flood. But Julia knew how much her daughter was looking forward to this party, and she also knew that if they postponed it, it would never happen. In short, the neighborhood was going to be flooded again, with all the chaos and disruption that would bring, so Julia moved the party up to that Thursday afternoon, before the storm. Some of the guests had already evacuated, and Julia's husband Jeremy had to work, but the party took place as planned—just a few days early.

Throughout the accelerated party preparations, Julia was also taking the time to go through the house and move things up off the floor. Closet shelves were stuffed full of important belongings. Bottom drawers were removed and placed on top of counters. But the important party prep still occurred, too:

> I mean I'm the world's worst Pinterest-Mom, but I was trying to, like, fake-Pinterest it up. I'd spent hours—seriously—doing a braid, then another braid, then another braid. We had made Leia-bun headbands for everybody. I'm terrible at all this. I don't enjoy this. But she wanted this so bad. Like so we had Jedi robes for everybody. I had Leia-buns for anybody who wanted them, which not one girl wanted them. Meg and I wore them, and this other mom wore them just to make me feel good.

Julia was absolutely determined to give her youngest daughter the party they had planned for months—complete with handmade party favors—even in the face of impending disaster. But this was not just about avoiding disappointing Meg. Julia was determined to proceed with the party because of what she knew loomed for her family. She wanted them to have this specific memory in their home before what would likely be a chaotic year. She was focused on what her family—and she herself—needed emotionally as well as logistically. In the end, aside from having fewer guests than originally planned, the moved-up party went off without a hitch, and that night at bedtime Meg proclaimed it "the best party ever." Julia was relieved. Jeremy thought the flood preparations and moving the party were overkill. They might be trapped in the house for a couple of days by floodwater, but this was all a lot of hassle over what would probably turn out to be nothing. Julia was thinking about Memorial Day 2015. At that time, they lived in a different Bayou Oaks home but were looking for something bigger, as they frequently had family come to stay with them. The Memorial Day flood came close to that house, but they didn't take on water. After the flood, sensing that deals might be available, Julia had had her realtor stalk new listings for her. With the realtor's help, Julia and Jeremy had found the perfect flooded five-bedroom home and bought it, embarking on a complete renovation of the interior before they moved in.

Why would anyone purchase a home that had recently flooded? As Julia explained, most people at the time thought Memorial Day was a once-in-a-lifetime storm. Bayou Oaks was a hot housing market, difficult to break into—even after the first flood. And a five-bedroom house in the neighborhood is rare, especially at a price Julia and Jeremy could afford. So it was the perfect house, in the perfect neighborhood, close to their friends, zoned to the same schools, and at a steep discount. It was a no-brainer. Boosting their confidence, the new house was not flooded in the Tax Day flood of 2016—a sign that many in the neighborhood interpreted as their being in the clear for future storms. Julia was not that sanguine, but she did feel more confidence in the home's prospects after that.

In late summer of 2017, after they had lived in the newly renovated home for only fourteen months, the bet the family had made was called as Harvey approached Houston. As Julia was preparing for the moved-up birthday party and placing the family's belongings on high closet shelves and counter tops, she called her realtor, who also happened to be her contractor:

> I said, "Get me on the list. I know everyone's gonna be calling. I'm on there." And he's like, "I think you need to have a more positive outlook about this. You didn't [flood on Tax Day], blah-blah-blah." I said, "That's great. I hear you. That's great. I'll keep my fingers crossed too. In the meantime, I'm on the list. I wanna know, can we live upstairs while we're doing the renovation? I wanna know do I call you first? Do I call the insurance? Talk me through some strategy." Give me strategy, right?

Julia was thinking three steps ahead—as usual. In addition to all of the preparations and party hosting that were in front of her immediately, she wanted to make plans for the inevitable recovery. She did not want her family to be left behind, scrambling to find a contractor along with the rest of the neighborhood. Next, Julia turned her attention to prepping for the storm. In addition to elevating furniture and stacking household goods on counters and tables, she prepped for a prolonged family stay in the house. Above the garage was a finished room with a half bath that they usually used as a playroom. In addition to bringing in a case of bottled water, she filled up every insulated water bottle they owned with tap water and lugged all of those upstairs. She packed five suitcases, one for each member of the family, with enough clothes for five days. She carted up snacks from the pantry, along with garbage bags, Ziplocs, and paper towels. She had her oldest son help her trundle the wine fridge upstairs. To her later dismay, she forgot to collect flashlights: "I was so smart about some things and such a dumdum about others—I thought we could use the light from our phones. And that lasted like two hours."

Julia's story of intensive preparations for the storm was common among the mothers. Although many first timers were taken by surprise by Harvey's severity, most of the mothers prepared for a possible flood, though the extent of those preparations varied based on whether the family had been flooded before. The prior two floods in the neighborhood meant there was wide awareness of practical preparations one could take to mitigate damage to the home and belongings, even among mothers who had not personally been flooded before. Families' preparations were improved by this *flood capital*, or the knowledge gained through prior flooding experience. Flood capital included things like knowledge of the approximate height to lift household items to (although in most cases this turned out to be too low during Harvey), awareness of which parking garages around town would accept cars to be parked up high, which FEMA and insurance numbers to call for the quickest response, which specific type of plastic storage bins would be watertight, which flooded household items could be cleaned and which would have to be discarded, and how to document and catalog ruined belongings after the flood. Mothers who expected to be flooded were not passive recipients of disaster; they proactively prepared to mitigate damage to their homes and families. These preparations fell into several broad categories: preparing the house and its belongings to take on water; preparing for the family to be trapped in the house for several days; preparing the children emotionally; negotiating with husbands about the extent of their preparations and evacuation plans; preparing for the inevitable postflood remodel; and preparing themselves socially and emotionally, enjoying one last wistful hangout with neighborhood friends before the flood.

With only a few exceptions, the mothers reported that *most of these preparations*—logistical, emotional, and social—were undertaken alone. Primarily, this is because mothers were falling into their usual roles as household logistics managers—albeit under unusual circumstances—but it was also because many of the husbands did not believe their homes were going to be flooded. Just as they benefited from the gendered division of labor in normal times, and from their wives' cognitive labor, or the labor involved in researching options, making decisions, and implementing plans, the husbands were (largely) outsourcing worry about the storm and attendant preparations to their wives.[1] That the mothers were expected to take on many of these tasks for the household has also been documented in other disaster studies.[2] In Bayou Oaks, many of the preparations were completed by women, either because their husbands were working or because the husbands did not believe the storm would be severe. In past studies, this has been explained by the fact that men are often working right up until the storm, so their wives, with more flexible schedules, take care of the preparations.[3] In this study, however, two-thirds of the mothers also worked outside the home, most of them full time. Nevertheless, in most cases much of this mental and physical labor still fell to them, perhaps because the husbands saw storm preparations as a natural extension of their wives' domestic responsibilities.[4] Whether the men were practicing "learned helplessness" or believed the preparations were unnecessary, or the mothers just considered their lack of help a continuation of a normal household dynamic, something that was annoying but typical, it was clear that in the end, the men benefited from their wives' preparatory labor.[5]

This chapter describes the lengths the mothers went to ahead of the storm to save what they could of their homes and possessions, as well as to mitigate the emotional impact on themselves and their children, often while also managing their husbands' emotions and unhelpful attitudes. They saw this labor as part of their mothering project—and as in Julia's story, their efforts reflected their intensive mothering styles. Just as they sought to curate their children's family, school, and neighborhood environments, they also strove to curate their family's hurricane and potential flooding experience. Before that curation could occur, the mothers had to assess their own household's risk of flooding, and they based that perception on a number of factors.

One measure of how much Bayou Oaks families felt their own homes were at risk was how much they prepared ahead of Harvey. Families' risk perceptions were based on prior flooding experiences in the neighborhood, on media accounts of the storm's possible damage, and on how their neighbors were reacting and preparing. This sense of risk was, in turn, amplified by the tight social network within the neighborhood. In other words, families' risk perception was socially constructed.[6] It was not based on precise weather models or scientific probabilities of flooding. Instead, mothers built their own risk forecasts based on prior experiences, but also by monitoring the feelings and actions within their social networks.

The mothers and their neighbors heavily based their perceptions of Harvey's flooding risk on the two most recent flood events: Memorial Day 2015 and Tax Day 2016. As a result, they underestimated both their projected risk of flooding and the level of water their homes would receive, often significantly. Because of the two recent flood events, the Bayou Oaks mothers believed that this upcoming flood would be similar to past floods they had experienced in terms of scope and scale. And to be "safe," they estimated that Harvey's floodwaters would be similar to the more severe prior flood, Memorial Day.[7] This belief that Harvey would be similar to the Memorial Day flood was so strong that nearly all mothers who had been flooded in 2015 prepped their homes for Harvey to precisely match the water level from that prior storm.

As sociologist Kathleen Tierney has pointed out, humans do not formulate their ideas about risks in a vacuum, but rather are influenced by a number of social and environmental factors.[8] In Bayou Oaks, even as they judged that Harvey would likely be similar to the Memorial Day flood, mothers adjusted that risk perception based on local media and on how their friends and neighbors were assessing and reacting to the possibility of flooding. In other words, although each Bayou Oaks family ultimately made their own judgment about how to prepare (or not) for Harvey, those decisions were strongly influenced by outside, collective factors. The mothers made decisions and took action based on a careful calibration of their own household's perceived flood risk.

The "social amplification of risk" framework illustrates how this process can operate.[9] In this model, "signals" about potential risks are amplified within social networks by institutions like the media. This amplification can both play up or downplay a risk, depending on how the signals are produced and interpreted. Mothers reported following the local news, although they also believed that local media were likely to overstate the risk from Harvey to get good ratings. Many of the mothers reported closely following a local independent meteorologist and journalist, Eric Berger, who had attracted an online following during previous storm events and promised "hype-free" forecasts. So when Berger started to get concerned about some outlier estimates of fifty inches of rain and posted these concerns on his website, this bubbled quickly through Bayou Oaks, amplifying and adjusting upward the level of risk the mothers perceived. Although signals were certainly coming from the media, in Bayou Oaks signals were also amplified by the residents' past experiences and exchanged between neighbors and friends. Actions were influenced by what others were doing to prepare, and this information was rapidly passed from one person to another via text message or Facebook. As Harvey churned ever closer to Houston, the pace of the signals accelerated.

Anna, a medical professional and mother of two who was flooded for the first time during Harvey, provides a powerful example of how this convergence can alter risk perception. Anna was at a conference in her California hometown when the storm approached. Although she likely would have done some preparations to her home if she'd been in Bayou Oaks, she was removed enough from the situation to not pay much attention until just before the storm. Two days before the storm hit, she was at her mother's home having dinner while they watched the national news:

> Friday I was standing in my mom's kitchen and we turned on the the NBC Nightly News, and [my neighbor] shows up on the news. So I'm like ... (a) I know them well, and (b) if my neighbors are on NBC Nightly News, the hurricane ... this does not bode well. So that's kinda like when I hit the panic button. When [my neighbors] are on NBC Nightly News, I'm like.... [stricken look] So ... so Friday night was bad.

Anna felt disbelief seeing her neighbors on the news, and this significantly increased her risk perception. Anna called her husband, an attorney and former marine ("which works really well for some occasions and it's terrible at other occasions"), and he lifted as much as he could in the home before evacuating with the children and his mother to an extended stay motel in Austin. Later, the family would be extremely grateful for this evacuation decision.

The individual risk perceptions differed considerably for families who had and had not previously been flooded. Most first-time flooders, seventeen of the thirty-six mothers, did not do any significant prep until it was obvious the water was going to come in. The few first timers who did undertake extensive preparations tended to be families who lived very close to the "water line"—the geographic boundary of flooded homes—from Memorial Day 2015. Based on past experiences, then, and with an eye toward Harvey's outlier predictions of rainfall, these families judged that their risk of flooding was high enough to take action ahead of Harvey. This was the case for Laura's family. When the waters had risen on Memorial Day, they had stopped just short of Laura's house. Laura watched the Harvey weather reports carefully, along with her husband Tony, who had grown up along the Gulf Coast in Louisiana. In fact, Laura and Tony had lived in Baton Rouge during Hurricane Katrina, so they had experience with a significant hurricane and its aftermath. Putting together what she was seeing on the weather reports with their own experience in Bayou Oaks over the previous two years, Laura began to feel more certain that this time they would take on water: "With how things were looking, it seemed like our luck might run out with this one."

Laura assessed the weather forecast and their past experience with watching neighbors be flooded, and talked with Tony. Having made the call as a team, she and Tony began lifting things inside their home. Laura had a big stack of clothes waiting to be donated to Goodwill and drove it to the donation center so they wouldn't have to deal with clearing those out after a flood. She took everything off the floors of their closets and stacked plastic bins on top of the guest bed, all the way to the ceiling. Then Tony went to Home Depot and picked up some cinder blocks, which they stacked to make risers for their precious family heirloom furniture items. These kinds of activities were typical of other first timers who did lots of preparation. Some kinds of prep were ultimately futile, such as sandbagging the house or sealing the doors with plastic. And some minor preparations—like propping furniture up on blocks or putting the legs of furniture into red plastic Solo cups—did no good in the end, as feet of water arrived instead of the expected inches. But the small group of first-time flooders who prepared extensively were nonetheless able to save a substantial amount of household goods by virtue of their preparations.

Most first timers, however, made only minimal preparations. The primary reason was that although Harvey was projected to be a bad storm, because these families had not been flooded in the prior two neighborhood floods, they felt a false sense of protection. In other words, their perceptions of the risk of flooding were altered *downward* by the prior two neighborhood floods, not *upward*. Having seen their neighborhood flood twice, and having been spared each time, produced a collective frame of reference that suggested their homes must be just far enough from the bayou, or elevated just enough, to be protected.[10] After all, the Memorial Day flood had been unprecedented, and they were spared from that. In the neighborhood lingo, these homes were called "Bayou Oaks unicorns." So while they anticipated their neighborhood would flood again, most of them believed they personally would not be flooded.

Mothers who made minimal preparations were usually like Amy, who at first told me that she and her husband didn't do anything to prepare, but then remembered a few specific actions she'd taken:

> Saturday night I did take all of our photo albums and put them on a folding table. Then Sunday morning I walked around and picked up the toys that were on the floor. We rolled up one carpet at the last minute. But all the other carpets, we sort of left 'em down. Like we put stuff on top of surfaces, but not nearly at the extent that I probably would've if I'd really, truly believed that the water was gonna come in. Not like the people who are like "this is what happens" and know.

Amy perceived the risk as slight enough that she only undertook a few precautions, many fewer than she would have if she had "really, truly believed" flooding would occur. Placing a few precious items, like wedding albums and baby photos, on the second story, in the attic, or on the top shelves of closets was a common minimal preparation among the first-time mothers. And some first-time flooded families did not do even this much, merely laying in liquor and wine and snacks to prepare for being trapped in their homes for a few days. These families were destined to frantically rush around as water flowed inside their homes, trying to save the most important items.

In contrast to the first timers, mothers who had been flooded before generally made extensive preparations as Harvey approached. I first interviewed Melissa, a 39-year-old brunette, at my dining table while our sons played video games in my den, just out of earshot. With a ready laugh, Melissa, a stay-at-home mom who also had been flooded in 2015, was resolutely cheerful even though her words made clear how emotionally difficult this experience had been. Her approach to preparations was typical among mothers who had experienced flooding before. The afternoon before Harvey hit, while her husband was at work, she walked from room to room in her home, taking a mental inventory. "That, I can clean," she thought. "That, I don't care if we have to get rid of it." She moved toward her dining room, taking in her grandparents' bookshelf. "That," she said to herself, "That we need to save." She disassembled it as best she could by herself and stacked the pieces on top of the dining room table. In her bedroom, she grabbed her hope chest, filled with years of memorabilia, and carried that to the dining table too. She removed all the lower drawers in her kitchen and bathroom cupboards and stacked them on top of the counters. She went to each bedroom and lifted all the clothes on lower racks in closets to the top racks. In all, she spent more than ten hours, alone, working to prepare her home. She felt a little silly; what if nothing happened? It would have been a lot of wasted effort. She thought to herself, "If me doing all this preparation causes nothing to happen, then I'm happy with that."

Lucy, a sunny, 35-year-old blond with three children, also prepared extensively for Harvey. She and her husband Michael had bought their Bayou Oaks home as-is after the Memorial Day flood, intending to renovate it. Before they had a chance to start on the renovation, the Tax Day flood hit, flooding the home yet again. Spooked, they briefly tried to sell the house, but felt the offers they received were too low. So they decided to proceed with the renovation and ride it out. As she put it, "We were like well, you know, let's just do it and then we'll see what happens." The renovation was completed in February 2017, and the family moved in, on edge when any rain event occurred. Early in the week that Harvey approached, Lucy was anxiously watching the weather reports and wanted to start preparing their home. She was stymied for the moment, however, because she was cochair of Bayou Oaks Elementary's upcoming Sno Cone Social, the traditional back-to-school, meet-the-teacher event. So instead of lifting household items, she was driving around borrowing coolers for the event from local families. It never even crossed her mind to reprioritize—school events were that critical to this community. This task *was* a good way, however, to check in on how her neighbors were feeling about the storm. While some of the mothers she saw, five days ahead of the storm, were not even aware it was coming, others were planning to prepare their homes but hadn't started yet. Eventually, the school's administration called off the event so that teachers could prepare their classrooms and homes for possible flooding. Lucy was relieved to have some extra time to prepare her home.

Once the event was canceled, with the knowledge of the two prior floods in mind, Lucy set to packing up their belongings. Like Laura and Tony, Lucy and Michael were on the same page, so he pitched in for the preparations. Michael took Thursday and Friday that week off of work, and the pair packed up an enormous amount of household items. They put a lot into their attic, but they took much more over to Lucy's mother's home, in a nearby neighborhood that had never been flooded. They filled her mother's garage with boxes and bins. Another mother, a friend of Lucy's, told me that Lucy had "packed up her entire house," which was only a slight exaggeration. In the end, nothing was left in the home that they couldn't live without: "And so we were able to save all of it. And that's why I always feel like the most fortunate, because we knew this was coming and we knew. But so many people had no clue and they lost all that."

Not all those who had been flooded previously were able to save this many household items. Several mothers prepared their homes by only lifting items above the water line from the prior flood, figuring that's what would happen again. Meghan, a 50-year-old mother of two teenagers, said: "[We] emptied every bottom drawer in every dresser and in every bathroom and everything, because the first flood we had 17 inches. So we figured okay that's good enough, right?" While prepping to the Memorial Day flood line wasn't enough for any of the Bayou Oaks households, at least these families were able to save some of their possessions. Some of the mothers focused just on saving essential items, worrying less about everything else. Rebecca, who had also been flooded on Memorial Day, tried to save her two favorite chairs this time by stacking them on the couch: "I put up my favorite chairs—which I've had to recover over and over again 'cause disaster keeps striking them—and I'm gonna be buried with these two fucking chairs, because I've spent billions of dollars at this point on them."

While some of the husbands helped with these sorts of preparations, as Laura's and Lucy's did, the preparations went well beyond lifting or packing away household items. There was additional anticipatory labor to do, looking ahead to the period after the flood they believed was coming. The experienced mothers knew they needed to be on the lookout for rental properties and storage units, as those would get snapped up quickly after the coming flood. These experienced flooders didn't wait to see if the flood would actually impact their homes. All of this anticipatory labor, involving tracking down rental homes and storage units, according to the mothers, was done by them. One mother not only booked a storage unit but knew to be prepared with her documentation when she showed up:

You know, having done this before, I was immediately at the storage place trying to get a storage locker. Actually I reserved it online the night before and then was there first thing in the morning, and he goes, "We don't think we have any," and I'm like, "Look, here's my confirmation, I reserved it last night." And we got one of the last ones.

I interviewed Meghan, a small business owner, for the first time in the dining room of her large rental home; the interior, to her dismay, looked like it had come from a 1980s time warp, but the house was in a great neighborhood and close to her two teenagers' friends. And most importantly, the rental house gave her teens space instead of everyone being crammed into a small apartment, as had happened the first time they were flooded. Warmhearted and friendly, but clearly also tough as nails, Meghan did not mess around when it came to finding a rental property. In fact, she had scouted potential places, including the one she chose, three days *before* Harvey actually hit: "Just in case, right? So I found a couple [nearby]. And I took pictures [screenshots] of [the listings]. So I knew what the rent was [before the storm]. Right?"

Meghan scouted potential properties and kept "receipts" by taking screenshots of the listings ahead of the storm. She took these precautions to avoid rent gouging later, something that several other mothers experienced when trying to find rental properties. Her family also benefited by the landlord being out of town (and thus out of the loop) during the storm, and by having substantial assets to draw upon: "We paid 12 months' rent up front. [Interviewer expresses surprise.] Yep. We had to have it wired. To keep the price what it was. This guy was out of town. I wanted him to sign like that before he figured out what was goin' on here." Meghan's experience illustrates how the mothers with flood experience acted quickly before the storm even hit and also took additional precautions to prevent being taken advantage of. Meghan researched available properties, noted the rent before the storm, and made sure to get the house she wanted at the price she wanted by quickly sending the landlord a year's rent.

Some other experienced flooders took extraordinary steps to protect their belongings and properties, again coordinated by the mothers. One mother, days ahead of Harvey, called a piano storage company and had them come take away and store their piano, at the cost of several hundred dollars. She also had wooden blocks cut at a home improvement store—to the precise height of their prior flood plus a few inches, in case this time was worse. Then she and her husband elevated all of their furniture on the blocks. Similarly, a mother who had been flooded twice before Harvey sandbagged her entire home and wrapped furniture and other items in massive quantities of plastic wrap, with the help of her husband and nephew.

Meghan's story, and those of others who undertook extensive preparations, are examples of *class-related efficiency*. Because of Meghan's access to resources and knowledge of how landlords operate, she was able to secure an advantageous temporary housing arrangement for her family, by searching before the storm hit. The Bayou Oaks mothers were able to tap into liquid assets for preparations, purchasing boxes and bins, buying or renting fans and dehumidifiers, having heirlooms moved into storage, and preemptively securing storage units. They were able to take days off of work to prepare without worrying about missing a paycheck or losing their jobs. The stay-at-home moms were able to rely on their husband's earnings while they focused on preparing. In other words, the mothers' assets, time, labor, and job autonomy were channeled before the storm such that they could save more items, and proactively protect their homes, more efficiently. This efficiency, enabled by socioeconomic position, as we will see, continued to benefit the Bayou Oaks mothers during the storm and after Harvey as well.

In addition to preparing their homes for a possible flood and gathering supplies for an extended homebound period, mothers were focused on protecting their children from anxiety and fear about the storm. The mothers put a lot of thought into how best to curate their children's storm experience, and they engaged in several different strategies that were all intended to prevent emotional trauma. About half of the mothers actively engaged their children in helping with the preparations, while the other half tried to keep the kids occupied with other things while they completed preparations. Although the strategies differed, the intent behind them was the same: to keep the children calm about the approaching storm and probable flood. Those who focused on keeping their kids preoccupied tended to have younger children and deemed them less likely to be helpful and less likely to understand what was happening. The rest of the mothers subscribed to a "helper" mentality for both practical and philosophical reasons. Practically, they were able to get more preparations done when they pressed their children into service. Philosophically, they believed that actively engaging their children would help them feel some agency over what was happening to them. Indeed, the work of sociologists Alice Fothergill and Lori Peek after Hurricane Katrina shows that not only can even young children be helpful in preparing for and in the aftermath of a hurricane, but giving them responsibility can increase their sense of self-efficacy and help them process emotionally what has happened to the family.[11]

Whether they chose distracting or helping as parental strategies during storm prep, almost all of the mothers also engaged in a third strategy: creating a sense of adventure and fun about the approaching storm. Mothers included in their preparation to-do lists obtaining new games, puzzles, and books for their children to play with, anticipating power outages and several days of boredom. They asked their children what sorts of snacks they'd like to have during the storm. They took advantage of a lull in the rain before the hurricane hit to send their children out for physical exercise and to see their neighborhood friends. And they framed bunking together in a hallway or closet during tornado warnings as an adventure, like a campout. Similar to the distracting and helping parenting philosophies, mothers believed framing the experience as something fun and unusual that the family would do together would help alleviate their children's fears and anxieties.

Laura's children, ages 11 and 8, assisted their mother by picking up their rooms and stacking items on beds and dressers. While they prepared for bed that night as the storm approached, Laura also briefed the children on what to do if and when their home flooded overnight: "We told the kids, 'We need you to get some good rest. If this happens, we're gonna wake you up pretty quickly and we've all got jobs.'" Laura believed that engaging her children in this preparatory work kept them from "freaking out" because they were busy. In addition, her instructions on what to do if water came into the home also served to calm her children's anxieties. The family had a plan—and everyone was on the same page. Their plan, with preassigned "jobs," would help them save more household items quickly and efficiently if water did come in. This was an intentional strategy Laura drew upon from her mothering repertoire; she knew her children drew comfort from having a plan.

Some of the older children even snapped into action without direction as it became clear their homes were going to flood. Tara, who taught fitness classes "for fun" in addition to her full-time job, had three children, ages 12, 6, and 4. As Tara, a first-time flooder, was preoccupied by preemptively filling out their FEMA application on her phone, Madison, her 12-year-old, quickly got the lay of the land and started lifting items in her room on top of her bed and dresser. Then she passed the word on to the little kids to do the same, barking out orders to them that Tara overheard: "Hey! If you've got anything on the floor that you wanna save and it's important, get it up high. Put it on your dresser!" Later, Tara's husband passed through the hallway, telling the kids to pack their suitcases with essentials:

> So the six-year-old goes and gets her Ariel luggage, flops it open on her bed, right? So she can start getting clothes out of her drawer. And her four-year-old brother takes his entire toy box and dumps it in her bag. And my husband's like, "Andrew!"—and he was like standing there going "What? You know . . . you said essential items, so I put all of my toys in there." So of course, she's like "Andrew, those are not essential items! He's talkin' about underwear, and pants, and shirts . . . not toys."

Even Tara's 6-year-old daughter was able to authoritatively advise her little brother on what to pack in his go-bag.

> Mothers who actively engaged their children in helping with preparations often reported how proud they were of how helpful their children had been. In nearly every case, instead of acting frightened or anxious when asked to help, the children had pitched in without complaint. In contrast to the experience of first-time flooders, for families whose homes had already been flooded once or even twice, preparing for water to come into their homes had become a routine. Take Jill, 41, mother of three children, who reported: "My kids are pretty used to this. So it's sad. They're used to . . . like when it starts raining hard, they go to their rooms and they pick their things up. Like even the baby does."

Similarly, Rebecca's 11-year-old son jumped into action without much explanation or prodding:

> I said, "We're gonna pack up the stuff in case we get water in the house. And I want you to, you know, look around your room, pick up stuff off the floor"—because the last one was such a surprise there was stuff under his bed and all that. And so he immediately was like "Okay, let me pick up that stuff." Like he . . . I mean he's 11 and there's not that much foresight really. But he was thinking I don't want my stuff to get ruined. So he really did start kinda picking up his room, putting the things he thinks are most important on his top bunk.

Rather than engaging their children as helpers, about half of the mothers decided to try to shield their children from storm preparations. In general, they believed their children would get too anxious if they understood the extent of what might happen. Better, they reasoned, to keep them preoccupied with other things while the adults readied the home. Mothers were, of course, genuinely concerned about their children's emotional states. But they also believed that anxious children would be more trouble to manage during the storm, and that they could get more done without the kids being underfoot. Jessica had two younger boys, one in early elementary school and one who was preschool aged. While her husband was at work and her kids were occupied by screens, she moved items upstairs bit by bit: "I kinda did it in phases so I wouldn't scare my kids. I didn't even really say anything to them until [the day before]. I kinda . . . I didn't want the anticipation to stress them out." Lucy knew that she had to get her home packed up ahead of the storm but was also worried about stressing out her kids:

> And so it was like I gotta . . . I gotta get it ready. But I didn't wanna freak the kids out at the same time, because, like, the front room is their toy room. And I was like, "I gotta get all of that packed up." But I . . . I wanna push that to the very end 'cause they're gonna be like, "Where are all the toys? What's up with this?"

Although some mothers tried to shield their children from the preparations, many children noticed anyway and asked why parents were buying supplies and moving household items. Mothers described giving age-appropriate and matter-of-fact descriptions of what might happen. Anita, mother of two and a prior flooder, decided on a direct approach when her daughter asked about their preparations:

> Since Memorial Day, like anytime there is uh . . . any talk of like rain, or storms, or you know lightning and stuff, she's always stressed. She was like, "Are we gonna flood? Are we gonna flood?" And so yeah, this time we did, you know, tell her, "Yeah, this might be the event." We . . . always kind of believe in being, you know, having an open conversation, and letting them know what's going on, what might happen. And we told her, you know, "Yeah, it looks like it's a bad storm. But we'll see. We're gonna be prepared, and we're gonna be—you know—all together."

Ruth also prided herself generally on being open and honest with her kids:

> We talk about everything. Nothing is off the table. So yes, I said, "There's a storm coming. We're not gonna be sticking around for it.

Mommy's getting help from our housekeeper to lift things. Well, do you want to help me lift your toys onto the bed?" You know, "Whatever. Whatever's important to you. Let's pack."

Julia, who had stocked her upstairs playroom as though for the coming apocalypse, didn't hide her preparations from the children, but she did obscure the purpose of them. She wanted to avoid increasing her children's anxiety ahead of the storm:

> I said, "Saturday night we're all gonna sleep up there. It's gonna be really fun." I probably glossed over why. They knew it was gonna rain or whatever. I don't think we really got into that, because two of the three are worriers. So I wanna prep them because surprise is bad. And I don't . . . I don't need them worrying on Tuesday about Saturday. Right? But I was saying hey. . . . I was trying to frame it. I had a feeling —best case scenario—we're upstairs for a while without power. Right? I mean I was trying to frame this into a this is not a scary situation that we don't know what we're doing and we're flying by the seat of our pants. We've got this. We're planning this family sleepover. It's gonna be really fun. So I just was trying to make this a fun, planned . . . adventure was the word I kept using, actually.

If the family was going to experience an "adventure," Julia was determined that it would be not just fun, but planned, just as she planned other elements of their lives.

This "fun adventure" approach to storm preparations was common among mothers, whether or not they were directly engaging their children about the coming storm. Amy didn't want her two youngest children sleeping in their usual room during the storm, because of the large oak tree just outside the window. So she moved them to sleep with their older sister in a different part of the house, telling them it would be a fun slumber party. Another mother took her boys to Game Stop to pick out some used video games for entertainment, as long as the power lasted. And several mothers purchased arts and crafts kits to keep their children entertained during the anticipated homebound period. All of these activities were designed to make being cooped up during the storm something that was not scary, and that might even be a little fun.

A handful of mothers actually sent their children away before the storm hit, to protect them from experiencing possible trauma and to make logistics simpler in case of flooding. Two divorced and repartnered mothers sent their children to stay with the children's fathers, who both lived in homes that had never flooded. Two other families opted to have grandparents in nearby cities drive over before the storm and pick up their children to take them out of the fray. About half of the families who had been flooded before, after completing their home preparations, went to stay with family or friends who they believed would not be flooded. The mothers who relocated either their children or the entire family did not want their children to experience watching their possessions and home be destroyed yet again. They also in many cases wanted to avoid being stuck in the neighborhood while roads were impassable. Lucy and her family, whose home had been flooded twice before they ever moved in, stayed with her mother in a nearby neighborhood:

> They thought it was such an adventure since it had rained Saturday morning and there was puddles all over. So we went out jumping in the puddles in the neighborhood. They loved spending the night at their grandma's, and so they thought it was even more cool that mom and dad were spending it too . . . 'cause it's usually like they're being babysat. So it was like this big slumber party to them.

These prestorm evacuations did not always result in safe harbor. Because of the magnitude of the flooding, several families who went to stay with friends or family ended up being flooded out of the place where they had gone to take shelter during the storm.

Something that nearly all the mothers had in common was that they reported they had taken on the majority of the preparations for the storm by themselves. While it is important to note that these accounts are based solely on the mothers' perceptions of the division of labor before the storm, what is most important is how the mothers *felt* and what they *remembered* about these events, because these perceptions shaped their feelings and actions. In many cases, according to the mothers, these preparations were not just undertaken alone, but also while their husbands derided them for overreacting. Even when husbands did help their wives, this help usually came toward the end of the process, as it became clear that the storm was for real, as for Julia and Jeremy. According to their wives, and similar to other disaster accounts, husbands tended to be reluctant participants in the storm preparations because they believed it to be a lot of unnecessary work when the entire storm might miss Houston.[12] Spouses did not necessarily agree about their level of risk, as evidenced by Deborah and her husband Peter, who had narrowly escaped the flooding on Memorial Day:

> So then in 2015 when my whole neighborhood was devastated, and we got out of it by inches, I said to my husband, "It's gonna happen again. We're going to flood. It's going to be a $300,000 day and I don't want that to happen to us. I'd really like to move before it happens." And he said, "It's not gonna happen. It didn't happen [this time]. It's never gonna happen again." And he—I don't want to make him sound like an idiot... he's a very thoughtful person—it's just that it seemed unprecedented at the time.

This type of disagreement between spouses, in which the wife was more concerned about Harvey than the husband, was a common narrative. It is important to note that for many of the families who had been flooded or very nearly flooded before, there was real anxiety around all rain events, which are frequent in Houston. On average, Houston receives rain every three days, much of it heavy, especially in the spring and summer months. So these discussions about whether to prepare for a possible flood took place routinely, not just as Harvey approached. The neighborhood was tired of responding to looming threats and living life constantly on edge. One would think, however, that the widespread prediction of an upcoming unprecedented rain event in Houston would have spurred a bit more cooperation from the men. Jessica pestered her husband all that week that they needed to start preparing: "I kept telling my husband we need to move our furniture. And he's like okay, I'm really busy at work . . . we'll do it. And we literally did it on Friday night. When the kids went to bed we moved everything, as much as we could, upstairs." Jessica wanted to move faster to move their furniture, but her husband wanted to wait until the last minute. Some of the negotiations were enhanced by some wifely cajoling. A first-time flooder, who had an injury and could not do much by herself, got her husband to help by placating him:

> [He] was like so annoyed with me. It was hilarious. He was just like. "Really?" He's like, "We're moving all this stuff over to move it back." And I'm like "Not too much." I go, "All of this stuff is gonna maybe take an hour." I was like, "Let's just do a few things here and there."

In hindsight, this mother found her husband's resistance to moving furniture amusing, given the feet of water that ultimately invaded their home, proving her right. Rebecca and Paul, who had been flooded during Memorial Day, were on the same page about needing to prepare the home, but not about when they would start working:

> It was a few days before the storm, I said, "We need to pack up." He's like . . . he said something about . . . you know he was kinda brushing it off. And I said, "Paul, look at the size of that storm. Everybody's agreeing. We need to pack up." He's like, "Well, we've got time." I go, "We really don't. This is gonna take longer than you think." And so I had to push him a little bit to get started earlier than he thought we needed to.

The mothers' negotiations with their husbands were usually successful, even if they didn't get started as early as they preferred.

In other cases, the couples were so used to the wife managing household logistics that negotiations weren't necessary, and the husbands just passively waited for instructions. When I asked Tara if she was generally on the same page with her husband during the preparations, she replied:

> Yes, but only because he generally lets me take the lead in situations like that. He's very good in that way. So we don't argue much about things like that. Like, he'll just say, "Like, what would you like for me to do?" And I just gave him that list of three things, and then he would come back and check in. . . . "I've done this, and this."

Tara's husband not only awaited instructions, but like a good employee he reported back on the list of tasks he was given. But he did not take the initiative to figure out what needed doing; he left that cognitive labor to his wife.

There were a few notable exceptions to this general pattern of wives being more concerned and prodding their husbands to take action before the storm. One mother of three young children followed her husband's lead on the preparations:

> I mean that week [my husband] was way more on top of it than I was. Like he was so worried that, like, you should see how much stuff we had in the house. Like food, water, you name it. And we even parked his car by my mom's house 'cause we had a feeling something might happen.

The typical gender roles for preparation were also reversed in Jill's household, with her husband Charlie pushing to get the house prepared and for the family to leave and stay with friends who had a two-story home. Jill didn't want to impose unnecessarily on their friends: "He thinks we're gonna flood every time it rains. So I really thought he was over-reacting. Yeah."

Unlike this handful of exceptions, the majority of mothers reported being in charge of storm preparations for their families. When I asked why their husbands were not more active drivers for the storm preparation, they would say things like, "It's just not his strength," or "He's used to me running things," or "I'm the logistics person." It's not that husbands didn't help at all; in fact, many did end up doing a lot of physical labor right before and after water started flowing into their homes. But the mothers, according to their accounts, were in charge of making nearly all the decisions, managing children, giving husbands instructions, deciding which items were most important to save, and evaluating where they would go once they evacuated. In "normal" times, the mothers intensively curated family life, and by extension, their family's status. And when an impending disaster loomed, the mothers also stepped up and redirected their managerial skills to optimize their families' outcomes.

Spousal conflict over storm preparations also bled into decisions about whether to stay in the home or evacuate to a safer place. Unlike during Hurricane Katrina, when most of the New Orleanians who had the means to evacuate left the city entirely, few Houstonians fled the city during Harvey, and city officials actually asked the public to stay put to avoid clogging the freeways, as had occurred during evacuation efforts before Hurricane Rita in 2005.[13] If Bayou Oaks families did leave home, they went to stay with nearby friends and family, or in hotels, in locations they believed from past experience would be safe from floodwaters. Nearly all of those who were flooded for the first time during Harvey stayed put, as most believed they would not be flooded; half of those who had been flooded before headed for higher ground, usually after considerable spousal bickering.

For many, the conflict was driven by the mother wanting to leave and the father insisting on staying.[14] Houston is far enough inland that wind damage was not a primary concern with Harvey, so families were making decisions based on the prospect of an anticipated foot or two of water coming into the home. The fathers generally believed that by staying in the home, they could mitigate damage as water began coming in. They also couldn't bear the thought of not knowing what was happening at their homes, and some were concerned about looting after the storm and wanted to protect their property. This attitude exasperated the mothers, who thought it would be safer to leave and wanted to keep the family together, like Jennifer:

> I guess he just wanted to stay hunkered down and just keep an eye on the house. It's like a man thing to do. I'd rather just, you know, take the kids and just hide in some place. And it turned out—I mean—we ended up being rescued. So at one point we were stranded and we tried to call for help and nobody was answering. So he apologized. He said, "Sorry, I should have listened to you and we should have evacuated before the storm hit." But it was too late at that point.

Jennifer excused her husband's determination to stay as being attributable to his gender, implicitly comparing it to the "woman" or "mother" thing to do, which is to remove her children from danger. But like many of the other mothers, Jennifer acquiesced in her husband's wishes, and the family endured the storm in their home. Sometimes the anticipated spousal conflict about evacuating meant that the conversation never happened at all, as was the case for Laura, who didn't even attempt to convince her Louisiana-native husband to leave the house:

> I knew there was no way that I was gonna convince [him] to leave, which we need a psychologist to explain that. But no, through... through his experiences I think—just growing up and storms and things—and it would take a lot—even more than what just happened—for him to leave.

Laura anticipated her husband's reticence to leave and avoided the conflict by not even raising the idea of evacuation, staying in the home, like Jennifer, against her better judgment.

Not all of the mothers went along with their husbands' plan for the family to ride out the storm in the home. Meghan correctly anticipated her husband's dismissive reaction to evacuation, so simply formed her own plan, which did not include him:

> I told my husband, "I'll be home [from work] by noon. I'm packin' up the kids. We're gettin' out of here." I said, "I've already talked to"—our

friend lives up the street here—I said, "I've talked to her. She said we can all come with the dog." I said, "But I know you're gonna think that I'm panicking so you can come or not." And he said, "Yeah, I think you're panicking." I said, "That's fine, but we're going."

Evacuation was one arena in which the mothers generally were forced to involve their husbands in their curation of the family's storm experience. One way to avoid spousal conflict about evacuation was to stay put; another was to separate the family. When husbands dug in about staying, wives acquiesced, even though, like Meghan, they sometimes took the children to safety and left him at the house by himself. Families who had experienced floodwaters coming into their homes before were not eager to repeat the experience, and they tended to evacuate to what they hoped was safer ground. Ultimately, most of the families wished they had evacuated their homes before the storm. Despite the mothers' best efforts at curating a calm storm experience for their children, the magnitude of the floodwaters would ultimately make that impossible.

As Saturday dawned, a steady rain fell, and the families who had prepared their homes for the storm were mostly finished with their work. Most of those who'd decided to leave the neighborhood for higher ground were gone. Instead of the daylong rainfall they expected, however, there was a lull that lasted all afternoon and into the evening, which encouraged people to get out and about. Some sent their children outside to play or ride bikes to get some exercise before being cooped up, and others took the opportunity to go out for lunch or dinner. Some were even feeling that Houston might have dodged a bullet. Harvey was now projected to make landfall well south of the city, much farther from Houston than anticipated. Maybe the rain that had come Saturday morning was all the city was likely to get, and all this work had been for nothing. Some families, though, were still anxiously watching the weather radar and could see the vast swath of storms that spun out from Harvey's center, creeping ever closer to Houston. This combination of uncertainty, anxiety, and boredom lent itself to an overall celebratory air, sometimes called a "hurrication" along the Gulf Coast.[15] Several of the mothers attended a "hurricane party" hosted by a local family, where they drank hurricane-themed cocktails from red Solo cups and focused on attempting to be cheerful while the rain began pounding down outside. Mary, mother of one and a repeat flooder, reported that "I've always wanted to renovate my house" was the dark humor mantra of the evening. The positive attitudes of earlier in the day began to fade as the streets filled with rushing water. Still, the ambience was one of defiant socializing, one last celebration of the neighborhood they loved so much and which most now expected to flood again. As Allison put it: "So I was picking up on this kinda vibe that's almost like a party. It's fun. Get your wine, and we're gonna hang out, we're gonna hunker down."

The advance warning that hurricanes allow sets this kind of disaster apart from others that arrive without warning. Businesses close early and events are canceled, allowing people free time to prepare. Residents largely heed public officials' warnings to stay off the roads and stay home, especially in Houston, where the roadways are actually designed to be part of the drainage system. But in Bayou Oaks, the calm before the storm had an added emotional weight. They suspected they would be flooded again—if not personally, then as a community—and they wanted to celebrate one last evening together before experiencing a communal trauma yet again. So some of the mothers reported feeling celebratory, even giddy, that evening. Others, though, like Mary, watched the weather radar with unease:

> So neighbors were you know drinkin' beer, drinkin' wine, goin' down the street. We were all . . . you know just kinda pokin' fun at the big storm that didn't come. . . . And I . . . I felt uneasy. They were jokin' around about it, but I was like, "It's . . . it's still not over people." You know? I was not feeling confident at all.

Mary had trouble getting into the party spirit and felt like her neighbors weren't taking things seriously. Similarly, Nicole was at one of the hurricane parties but felt uneasy and didn't drink:

You need to be aware and alert, every time there's a storm. You know? So . . . I didn't drink that night because I was like the last thing I want is to have a hangover or be—you know—drunk, when the kids are scared, or when you know a tree falls on the house, or you know, whatever.

Nicole tied her lack of party spirit to the impending storm, but also to her mothering. She wanted to be ready to act if necessary.

In contrast to the socializing spirit that pervaded the neighborhood, several mothers just wanted to be home with their families. They weren't feeling up to seeing people or uneasily celebrating. Instead, they spent the time preparing themselves emotionally for what was to come. Exhausted from all of her preparatory work, and with household items piled on top of every surface, Melissa asked her husband to pick up Chick-fil-A for dinner on his way home from work. As her family sat and ate together that evening, chatting amiably, Melissa was quiet. After her kids were out of earshot, she said to her husband, "This was the last time we'll ever be eating dinner like this in this house." Melissa saw the writing on the wall. Although she had spent more than ten hours preparing, they would be flooded again, and their home would never be the same. That sobering realization was shared by some of the other mothers in Bayou Oaks that evening. Allison played board games with her young teenagers and husband on the living room floor that night, just to keep the mood light: "I think at some level, we thought this could be our last night in this house. . . . [I]t was sad, but we laughed and we had a good time."

Jennifer, who had only completed renovations from the last flood a couple of months earlier, remembered defiantly reveling in the feel of her new wood floors on her bare feet, "because I know it's gonna be gone, so I just kept walking on it without shoes on." She was determined to remember the feel of her new wood floors. Another mother cried when her daughter fell asleep before she could read her a bedtime story. "I remember knowing that that night was the last night in our house. And I'm not gonna get to read to her one last time in her bed." Reading a bedtime story was an important way this mother connected to her home and to her daughter. It was crushing to realize that was probably not going to happen for the foreseeable future.

The Bayou Oaks mothers who expected to be flooded mourned their homes *before* they were ravaged by Harvey's stormwaters. They mourned not just for the structure, but also for their carefully curated family lives, which they knew were about to become chaotic. There was only so much curating they could do of their families' storm experience; at best, they could keep their kids calm and save some items. But they could not save their homes. Their past experience with flooding in the neighborhood told them it might be six to nine months before they could be back in their homes, or much longer if they would have to build a new home. Parenting to ensure family stability is challenging enough without throwing a ruined home into the mix. However, rather than relax their expectations about mothering, as we will see, the mothers strove to keep family life as normal as possible, which meant continuing to exemplify intensive mothering ideals while living a displaced life. In some ways, this emotion work before the storm hit likely helped them afterward, when they would be overwhelmed with recovery and setting up a new, temporary home for their families. These poignant preflood images also represented family goals to get back to as soon as possible.

When I interviewed Julia again a year after Harvey, we sat in the living room of her renovated home, a stylish mid-century modern with a capacious great room that included the dining room, an open kitchen, and the living room under a vaulted ceiling. A well-outfitted wet bar hugged the wall near the sliding doors out to the backyard. Julia and Jeremy were back in their home within four months, moving back in on Christmas Day 2017. As soon as we sat down, I told her that they were the family furthest along in the renovation process in my study. "I'm the most, I'm the most furthest along family you've—in, in the world. Not just that you—I—that—I have come in contact with—I'm definitely the furthest along. I think I shattered all the statistics."

Julia took pride in how quickly her family was back in their home and back to their typical family routines. The personal agency expressed in Julia's statement was common among the mothers. And I knew exactly what she meant by "I shattered all the statistics"—because as she would describe in detail, she alone had managed the entire process, start to finish. Moreover, she attributed the speed with which they had returned to the neighborhood to her own extensive preparations before the storm. When I expressed my amazement that they were back in just four months, she replied:

> It is, but not if you think about why. I mean, I—because I prepared for four days. I mean, I had all the drawers—I saved us so much time in terms of carpentry because I had kept every single drawer, so they didn't have to come back in and remake the drawers, they just built the boxes again around the existing drawers. I had put tile in in the first place, so that I—we didn't have to redo the floors. So, there were zero decisions to be made because I went back with the same thing I had last time, so I didn't have to go shopping for fixtures, or think about—paint colors. It was as streamlined as humanly possible. I mean, I, I, I had it all planned out before it even happened.

The mothers of Bayou Oaks had seen this all before, even if they had not personally been flooded before. Most of them prepared—some extensively—for the floodwaters that would ravage their homes. Those preparations were based on their own perceptions of risk that were influenced by their own past experiences, media reports, and their social circle. While many of the mothers' specific preparations for the storm were overwhelmed by the volume of water that ultimately came, their actions did save some items. More importantly, the preparations gave mothers a sense of control and agency during a time when they otherwise could have felt helpless. Just as they curated their families' daily lives, they sought to curate their storm experience. The typical gendered division of labor in their households, with the mothers responsible for most of the decision-making and logistical arrangements, largely worked as intended during preparations for Harvey. The mothers' skills at curating family life were temporarily redirected to storm preparation. This intense responsibility, with its attendant mental and emotional load, however, would have long-lasting consequences for these mothers. It also would not end once the storm hit, but would continue during the flood itself and throughout the next year.

3 During the Storm

GET THESE BABIES OUT OF THE WATER

Amy, 36, a part-time freelance writer with springy dark curls, put her three small children to bed together in the same bedroom as Harvey approached. She could tell the kids were keying off her anxiety that evening, so she wanted to treat it like a slumber party in hopes that it would seem like an adventure. Although their home had not flooded in the prior two flood events, Amy closely tracked the weather reports and slept fitfully. She woke up at about 6:00 a.m. and immediately peered outside to see the water level. It was high, much higher than she'd ever seen before. Her stomach clenched. She stood and watched as the water crept up to the bottom of the swing in their front yard. A switch of certainty flipped inside her. They were going to be flooded. There was no point in watching and angsting. They should act.

The kids were uncharacteristically sleeping in, which Amy was glad for. But she decided she had to wake up her husband Mark. "I think water might come in this time," she told him. "There are things we can do, do you want to do them?" Amy and Mark briefly discussed trying to save their couches, but Amy dismissed them as old and "full of Cheerios anyway," so not worth saving. After lifting some important papers and mementos and saving a couple of rugs, Amy sent a text to their neighbors, who had a second story. "The water's getting really high out here. If we start flooding, can we come over?" The neighbors replied, "Of course."

Amy and Mark continued making preparations and got the kids up and dressed, just in case. Amy checked the water level outside again. It was nearly up to their porch. She felt panic rise again and forced it down. What should they do now, to prepare for possibly having to leave the house? She packed a go-bag, but then realized the wheeled suitcase would not be much use in a flood. She took another look at the water level outside and decided to make everyone breakfast—a real one, with bacon and eggs. Later, she would regret letting those breakfast dishes sit in the sink.

After breakfast, Amy told the kids to stay on the couch and turned on a movie. The kids were thrilled: they didn't ordinarily get screen time in the morning. The kids' excitement was in stark contrast to Amy and Mark's panic, which rose in sync with the flood outside. As water started audibly lapping over the slab of the front porch, Amy decided it was time to talk to the kids about what was about to happen. "We're going to get some water in the house," she told them. "It's OK, and we have a plan, but water will come inside. We'll go over to [the neighbors'] house." Amy believed that calmly talking to the kids about what was happening was critical—not only to address their fears, but also to let them know that everything would be OK because the family had a plan. She had some experience with anxiety herself and knew that her children would need the structure of a plan:

> 'Cause they're very . . . I mean like my mom says, I installed those buttons. You know? And so the kids actually like . . . it's not that we live a very structured life, but they like knowing what is next. And I like knowing the agenda for the day, so they like knowing the agenda for the day. So telling them these are the things we're gonna do, I think made it just like okay.

At about 10:00 a.m., water started flowing into their home from all directions at once. Amy had imagined that it would come in through the doors first, but it bubbled up through the slab foundation across the whole house at the same time. It wasn't as fast as she had imagined, but it was relentless. When the water was about an inch deep, she and Mark got the kids into their rain boots. Then, as they prepared to leave for next door, they stood in a circle and prayed. Amy wept as she described the scene to me: "We all held hands and we said . . . this house is not our home. We're our home. This is the important part."

As they left the house, wading through the knee- to thigh-high water in their backyard to the neighbor's house, with Mark carrying two kids and Amy carrying one, Amy saw something she'd never seen before. Bugs and lizards, dozens of them, were swarming up the side of the house, climbing as high as they could to save themselves. She was amazed at the mass critter exodus, but she found some commonality with them:

> I saw a spider mama with her egg sac. Her egg sac is like tied to her back and she's climbing up the stairs—and I'm not a fan of spiders—but I was looking at it, and I'm holding [my youngest daughter], and I was like, "I get you. I get you. We're both doin' the same thing. Get these babies out of the water." You know?

Amy's story illustrates how the mothers took charge during the flood. It was clear that they saw themselves as chiefly responsible for their families' well-being. They prepared their homes; they made the evacuation plans; they called on their friends and neighbors for help using the social networks they had created and sustained. And throughout the flood, the mothers weren't just making plans to get their kids to safety; they were also highly concerned with their emotional needs. They went out of their way to give their children the illusion of calm and safety. The mothers were used to being in control, and the floodwaters deeply threatened their identities as the guardians of stability for their families. Instead of surrendering to the situation, the mothers did everything they could to maintain that control as long as they could. The mothers, just as they curated family life, were determined to curate their family's flood experience to minimize the damage, both to their homes and to their children.

All across Bayou Oaks that Sunday morning, homes were taking on water, and families like Amy's were frantically trying to save their belongings and keep themselves and their kids safe. Those who evacuated before the flood to stay with friends and family felt helpless as news of the flooding reached them through social media or text messages. Some even watched the water rise inside their homes via security cameras they'd set up before leaving. Melissa's husband John rigged a doorbell camera to take periodic photos of the view from their kitchen into the living room. They used a bench by the kitchen window to gauge how high the water was getting. Image by image, the water climbed the legs of the bench. After a couple of hours, the seat of the bench was completely submerged. In the next image, the bench could no longer be seen, and in the next one, it was jauntily floating by in front of the camera. Melissa and John managed to have a dark sense of humor about the bench's adventures. As Melissa put it, "So it was funny—but like, we were not laughing like, 'ha-ha.'"

The mothers who had remained in Bayou Oaks were grappling with the damage to their homes and possessions, but also with the water itself, which they described as "disgusting." This was not clear rainwater that was rushing into their homes. Instead, brown, muddy, fetid "poop water" was bubbling up through the floors with a sound like "a witch's cauldron," smelling of sewage and chemicals and filled with unidentified debris. The memory of the water's odor stuck with nearly all the mothers. Rebecca said, "It definitely is—especially when it's fresh—it's a very acidic smell that definitely feels like it's burning your nose." Many of the mothers tried to minimize their children's exposure to the water, like Jennifer, who said: "'Cause you know it's dirty. 'Cause you kinda hear about and you just don't know what is in the water. It could be snakes. We just didn't know what was in there."

Mothers of younger children had particular difficulty keeping their children out of the water, but also felt it was urgent to do so:

> I don't know this, but in my mind [the water is] like combined with sewage, and dirt, and germs, and snakes. I mean it looks . . . it is, like, filthy. It looks brown . . . dark, dark brown. It just looked so dirty. And yeah . . . yeah, that's definitely a big thing of you know don't . . . don't touch it. You know they'd be like, "Oh look, there's a Lego floating." And I'm like, "Don't touch the floating Legos!"

Likewise Jill, on the second story of her friend's flooding home, was continually shooing her kids away from the stairs as they tried to creep down and look at—or worse, touch—the water: "And because my kids still put their fingers in their mouths. I can't get them to stop. Like, 'No dysentery please.' Like that's what we need on top of a flood."

The mothers were right to be concerned about what was in the water. The most immediate danger was the huge mounds of fire ants, which floated menacingly on top of the water like arthropod life rafts, but the moms' concerns about chemical and bacterial exposures were also rational. In all, the Associated Press found that more than one hundred toxic releases occurred before, during, and after Harvey, spewing carcinogens into the air and water.[1] While scientific studies of these releases' effects on individuals' health are still ongoing, early evidence suggests a wide array of both acute and chronic health impacts.[2] Texas's notoriously lax environmental regulations, coupled with a robust petrochemical industry situated along the Gulf Coast, make Houston especially vulnerable to toxic emissions and spills during hurricanes, both "accidental" and not.[3] During floods, industrial contamination is compounded by human waste; all over Bayou Oaks, residents and their pets were forced to relieve themselves in the floodwater as toilets stopped flushing or were submerged. Flood capital gained from prior floods included information about health hazards that might result from contact with the water. After the Memorial Day flood, several residents had contracted dysentery—something they hadn't heard about since playing the computer game Oregon Trail as children in the 1980s. Laura's husband Tony got a staph infection on his leg after helping with recovery after the Memorial Day flood, and stories of rashes and unexplained or exacerbated respiratory problems resulting from contact with floodwaters were common in the neighborhood. All of the mothers, then, were eager to get their families safely out of the flood waters once they arrived.

Everyone reacted differently once the water breached the house. Some were resigned to the damage and immediately left for neighbors' second stories. Others were determined to get a start on recovery even before water was all the way in. In one home that had never previously flooded, Nicole's husband Dave began furiously ripping up dry carpet as water started bubbling up through cracks in the walls and fireplace and carrying it out to the front porch. Nicole, who would have preferred in that moment to be saving furniture and household goods, described him as being "in his own panicky place" and decided to leave him alone. Later, though, she realized that he had been right, as carrying out strips of dry carpet was much easier than dealing with soaking, heavy carpet later.

Many families were concerned about electrocution, or frying their electronics, and endeavored to turn the power off to the house after water came in. Though at the time, Nicole was not amused, she later laughed as she recalled her husband's pause in the midst of the chaos:

> So I heard this buzzing . . . this electrical buzzing sound, while I'm sitting on the bed. And I was like, "Oh my god we've got to turn this electricity off." And so I sloshed back down the hall and I hear—this is so stupid—I hear the Nespresso going. I don't know if you've got one of those but they're really loud. And I said, "Dave, you have to turn the electricity off. I can hear it. It's buzzing. You know, something's gonna catch on fire." And he's like, "I gotta make this cup of coffee." And I just

stood there and stared at him and he turned around and looked at me, and like started drinking the coffee with his waders on. It was such a bizarre moment.

Dave wasn't the only one who needed a little normalcy as the water rushed in. One mother who had been flooded twice and then moved to a home she believed would be safe reacted to her new home flooding with defiance, determined to cook breakfast for her still-sleeping family while water soaked up the legs of her pajama pants: "I went downstairs and cooked a meal . . . in the water. Which is just dumb. Because there still could have been something electrical going on. But I remember scrambling eggs—water up to my ankles. I was like, 'I'm not letting it stop me this time.'" This mother wasn't having it. Her family had been flooded too many times. She was thinking about her family's needs at that moment, instead of trying to deal with the damage.

There were even some moments of levity. As Tara and her husband and three children were packing go-bags and preparing to leave the house, something they had hidden away from the children suddenly reappeared, casually floating down the hallway:

> So here comes the elf-on-the-shelf and I swear he had his hands up like he was floating in the pool. So elf-on-the-shelf is relaxing. And I'm sitting there going, "Where did he come from?!" And my husband started cracking up, 'cause he had hidden him in his bottom drawer. So somehow he magically floated out. 'Cause he's like, "I hate that!" And there he is floating down the hall. The kids are going oh—you know—they're so excited. They named him Sprinkles. They hadn't seen him in forever. And so I was just like okay, this is funny right?

This totem of middle-class self-regulation and consumerism, a part of many Christian and even secular mothers' December parenting structures, suddenly became absurd in the face of floodwaters. Nonetheless, all over the neighborhood, mothers were pushing down their own anxieties and modeling that self-regulation in hopes of keeping their children calm.

As the mothers watched their carefully curated home life start to come undone, they also took on the responsibility for making sure kids were emotionally as well as physically safe. This wasn't unusual; this was an everyday responsibility they carried—but the sense was heightened during the flood. The intensive mothering ethos was so strong that they automatically turned to it during a crisis. First, they reassured their children that they would keep them safe. They explained their plans so everyone would be on the same page. And they presented the flood's challenges as learning opportunities. Amy reassured her three children by explicitly placing the responsibility for their safety on herself: "One of the things I say to the kids all the time is, 'What's mommy's number one job?' And they're like, 'To keep us safe.' 'So we're gonna keep you safe and that's what's important. And this is how we're gonna do that.'"

Amy's mantra was one she had found useful in a variety of parenting situations, but it also demonstrates the level of responsibility she placed on herself. Her identity as a mother was wrapped up in keeping her children safe; it was her "number one job," something she'd even instructed her children to repeat. Leah echoed this sentiment, and I asked her why she felt that she was the arbiter of what the family should do during the flood. She framed her decisions during the entire flood experience as being about the children: "I think with a lot of the parenting stuff, you know, I'm more the lead parent. And [my flood experience] was very much about, like, what is best and okay for them."

Leah, like many of the mothers, saw herself as the "lead parent," and her husband agreed. Naturally, then, she would make the decisions about what the family should do during a crisis. The mothers accepted the responsibility to make decisions and plans and also to communicate those details to their children to help alleviate their anxiety. Across town, Melissa and her family were safe at her father's house, but her two children, ages 6 and 12, had lots of questions about what the family would do now that their home had flooded:

And I just kept telling them, you know, it's like, "We're gonna take care of everything. You don't need to worry about it." I've always told them that if something were to happen to us, our families would like, let us, you know, move in until we got back up on our feet. "You don't need to worry about that. All these people would help us. You don't need to worry about it. It's gonna be fine."

Even though in that moment Melissa didn't know exactly what was going to happen, she knew that her extended family would be there for them if they needed it. And she explicitly communicated that to her children, in an attempt to calm their anxiety. Like Amy, several mothers reported easing their children's flood anxieties by letting them know what the plan was, and what to expect. For many, this was already an everyday parenting strategy. As Tara put it:

Any time we're approaching something new, I try to give them the things that they can expect, because I know that that will calm their nerves. You know? Like, "Here are the things that we know about the situation that you're walking into. And you're gonna meet some unexpected things, but if you can just focus on those things that you know, everything will be okay." You know?

Another parenting strategy the mothers used was to focus on the practical aspects of what needed doing, so their children could see them working to address the situation. They saw this as an educational opportunity—a chance for their children to learn how to manage difficult circumstances. Rebecca saw the flood as a way to model for her older son how one ought to react in a crisis:

I hope that he was just kinda listening to what mommy was tellin' him to do. And I don't think I was in any way panicking, at least not externally. Because I can tell . . . he's so tuned into my emotions, which . . . number one, there's the adrenaline of being in the situation. There's just the way I work in general, which is there's a problem. So let's analyze it, and let's solve it, and let's do everything in the right steps.

Part of the intensive mothering ethos is that there are, indeed, "right steps" to take in any situation. Parenting during a flood was no different. In addition to making decisions that focused on the children's needs, communicating plans to the children, and modeling problem-solving, the mothers worked hard to stay calm and, more importantly, to demonstrate to their children that they were calm. For many, this involved intensive personal emotion management under difficult circumstances. Inwardly, the mothers were dealing not only with knowing their homes and belongings were ruined, but also with the knowledge that the family life they'd so carefully constructed, and which they valued so much, was going to be disrupted for the foreseeable future. Nonetheless, most of them worked hard to present a calm aura of unconcern to their children. Julia's kids, stationed upstairs, didn't even know it had flooded downstairs until she told them:

> So I said, "There's a little bit of water downstairs, but, you know, we'll deal with it later. Don't worry about it." It was super important to me the way we frame all this to them. I kept saying to Jeremy, like, "If we can just be really calm...." " I mean I know kids. Right? I'm a kid professional. They're looking to us [for] how to respond. If we're calm ... if we're prepared ... if we show them that we're okay, they're gonna be okay.

Julia downplayed the amount of water—"a little bit" was nearly three feet deep—but she told her children not to worry because she knew she'd handle it. And while her own heart was racing and her palms were sweating with worry, she believed her kids would be better off if she could hide that and project calm. Despite the mothers' best intentions, many kids nonetheless noticed their mothers' managerial efforts—especially the older ones. When Tara praised her children for following directions and staying calm during the flood, she got back something she wasn't expecting:

> But then [my older daughter] said to me, "Mom, I also think that it was because you stayed calm." And she's like, "I think if we had seen you break down and cry ... like you always told us if you were crying, or if someone was yelling, that's when things got bad." And she's like,

"So we didn't think anything was bad because you didn't cry." Right . . . you know? Like, that's awesome. You know my husband said the same thing. He was like, "Well, I was just, like, reading your temperature." And I was like huh . . . so mom was responsible for everyone's emotions at that point. Right? Hmm . . . .

After learning that her children and even her husband had been looking to her for how to react, Tara realized the importance of her own emotional regulation for the whole family. She clearly viewed that fact as an unfair burden—one that only she carried. And unlike Amy, who explicitly told her children she was responsible for their well-being, Tara was ambivalent about bearing that responsibility.

Some of the mothers regulated themselves only after noticing their anxiety was impacting their children. Nicole was frantically trying to save toys and books in her young daughter's room from the floodwaters when she realized that it would be better for her to just chill out:

And then I just looked around and I was like, I can't save all this stuff. And I just stopped. 'Cause I saw that she was starting to panic. And so . . . we sat on the bed and we got these sticker books that you . . . make. It looks like a painting at the end. And we started doin' that. And I just kept lookin' down the hall and I remember seein' the water like . . . you know the carpet would start getting darker and darker, you know, with the wet. And I would just like ignore it. I would just totally . . . I just . . . I just ignored it.

Nicole focused on her daughter and ignored the creeping water, which helped her to regulate her emotions. Molly, a three-time flooder and stay-at-home mother, learned emotional management from her experiences during the first two floods:

For the first flood, you know, we were all stressed out. So the second time . . . the second time was a little smoother 'cause I was a lot . . . they weren't feeding off of my emotions. So and then this [third] time around I'm very calm. And yeah, and the girls are . . . they've done it twice so I guess it's like, "Mom and Dad will handle it."

Molly's girls benefited from their mother's prior experience managing her emotions; for them, this third time was annoying but not a huge deal, because their mother was not outwardly stressed and they knew they could count on their parents to manage the recovery.

There were certainly some exceptions to mothers hiding, or managing, their own emotions with their children. Interestingly, this strategy was also framed as being for the children's benefit. One mother of two said she "just bawled" as the water continued to rise. But in her view, this was a good thing, not something to be hidden:

> And my husband and I talked about this, and my kids were never scared during this whole ordeal. Uncomfortable? Yes. Rained on, soaked through? Yes. Frustrated that things were taking so long and where are we gonna go? All of the above. But they were never scared . . . probably the only thing I can think of is maybe we were just matter-of-fact about it all enough. Because my initial emotional state was something that simply needs to happen now, not in two hours. I need to . . . to move on immediately. I need . . . and so that's when we did. You know? I had a few tears and we moved on.

This mother believed that her general good attitude during the recovery process was enhanced by allowing herself to feel grief right away, even in front of her children. Her instincts may have been correct. All of that emotional self-regulation the mothers engaged in for their children's benefit had a cost. While most of the mothers explicitly did not give in to their emotions because they believed it would upset their kids, several expressed regret about that in their initial interviews. Many of them believed that suppressing their initial emotional responses had caused them emotional and physical harm in the weeks that followed the storm, as stress-related health impacts began to surface. The mothers also thought it was possible, in hindsight, that hiding their emotions wasn't a good example for their children. Ruth believed that while managing emotions was helpful for kids to a degree, it was also good to let them know it was OK to feel sad: "Clearly we set the stage for how our children feel about things. And I'm keenly aware of that, so I've done my best. But there are times when I've also cried and talked to them about being sad, you know. So [I felt] the gamut of emotions for sure."

The mothers were feeling their own sensations of rage, fear, disbelief, and grief. But with considerable emotional labor, they modeled calm, rational decision-making under pressure for their children because they believed that to be what their children needed. And even when mothers chose instead to allow their emotions to show, they also framed that as best for their children. This emotional labor, which would ultimately last for weeks, was yet another kind of labor these women undertook as a result of Harvey. Some of the women told me during the initial interviews that it was the first time they had allowed themselves to feel emotional about what happened to them—there had simply been too much else to do. For these women, accustomed to managing enormous mental loads as they curated their family's daily lives, this process of regulating their own reactions and feelings was nothing new. Nor was managing what they considered the appropriate parenting response, something they did on a daily basis. But the fact that they were able to do both during a literal disaster reveals the strength of the child-centered, intensive mothering ethos.

Bayou Oaks in 2017 was dominated by one-story homes, with only one or two two-story homes on most streets. As floodwater pushed families out of one-story homes, many of them were forced to converge on these scattered two-story homes in uncertain circumstances, sometimes with strangers and often with frightened animals. For many, it was deeply uncomfortable—an invasion of privacy—to wade into someone's flooded home and track mud and detritus upstairs. Without power and thus air conditioning, late August in Houston is unbearable. The swampy air was thick with the smell of sweaty people and wet dogs, and soon, with pet excrement. Add a feeling of being trapped in a small space for who knows how long—and not knowing how far the water was going to rise—and you have a recipe for stress. In this next phase of the flood, seeking safety and shelter with neighbors, the mothers tried to regain control of the situation by either enacting the host role, if they were sharing their own homes with neighbors, or enacting the role of a good guest. And despite the disaster setting, they tried to engage their children in enacting the hospitality norms as well. Even in these dire circumstances, with tempers fraying and stress levels high, the mothers were worried about their children's behavior. They worried about imposing on the people hosting them. And the hosts had to deal not only with their homes flooding, but also with a mass of dirty, wet people and animals being forced inside and having to share their resources. Once out of their own domains, the mothers could no longer control what their children saw or heard about the flood. Their ability to curate the experience was essentially gone. They could not protect them emotionally in the ways that they wanted to and had to redouble their efforts to keep up the sense of fun and adventure.

Shannon and Sean, both architects, had added a raised, five-hundred-square-foot addition to the back of their modest ranch home after buying the house, in what was meant to be just a home improvement but ended up being a stroke of foresight. Raised about four feet above the ground, connected to the rest of the house by a small set of stairs, the new media room connected via a spiral staircase to a new master suite up above. As their home began to flood, they retreated along with their two girls, ages 10 and 13, and the family pets, to the addition. Before too long, the neighbors behind them, Laura and Tony, texted to say the water was getting too deep inside their home and to ask if they could send their two children over. The journey to Shannon and Sean's—the house directly behind theirs, separated by a privacy fence—was perilous. Going the long way around using the street wasn't an option, as the water there was neck deep and the current was too strong. As Laura prepared to send her kids out into the water, she fretted. She had been holding it together so far, but the water was so deep in their backyard by this point—well over her thighs—that she feared for her children's safety. She got her children, ages 8 and 10, dressed in their raincoats and rain boots. They put on their backpacks, and then each parent carried a child piggyback to the fence. Tony placed a ladder on one side of the fence, Sean placed a ladder on the other side, and they carefully got the two children up and over.

Around this time, another neighbor, Ellen, mother of two tweens, was huddled on her bed trying to calm a large standard poodle who was terrified of the floodwater flowing through the room. She and her husband Bill decided it was time to get the kids to safety, so Ellen texted Laura to see what they were doing. Laura let her know that they had sent their kids over to Sean and Shannon's, and although Ellen didn't know them at all, her husband, with Tony's help, got their kids over their own fence to the relative safety of the elevated addition. Eventually, Laura decided to wade over to Sean and Shannon's too, but could not convince Tony to come with her. Although before the storm she hadn't tried to argue with him when he didn't want to evacuate, by this point she was weary of his stubbornness: "I was just worried. You know? Bulldoze the house down right now, that's fine. We need you. We don't need stuff. We just . . . we just need you."

Frustrated and feeling a powerful instinct to be with her children, Laura left Tony at the house and made it over to Sean and Shannon's. After Ellen saved a few more pieces of art on her home's walls, she found a poncho and waded over to Shannon and Sean's as well. Meanwhile, Bill got the family kayak—a fortuitous wedding gift that proved to be unexpectedly valuable over the years in Bayou Oaks—out of the garage and floated it into the house and down the hall to the master bedroom to rescue the dog. Somehow Bill managed to get the terrified dog into the kayak and paddle him out of the house and around the block to Shannon and Sean's.

Inside, Shannon's inner hostess was taking over. She organized the group—now six adults and six children, two dogs, and one cat—so that the kids were upstairs in the master bedroom and the grownups were downstairs in the media room. The small room just had a couch and a couple of chairs, so people sat wherever they could find a place. I asked her what the mood was like: "Well, the parents were quiet and just staring. The kids were doing their best. What was nice was I just sent them upstairs. I stabilized upstairs and we cleaned some space out."

Just as the group started to relax a little, they heard a knock at the door. It was Jim, from the house directly next door. None of them knew him, because he and his wife and three young children had just moved into the neighborhood twelve days earlier. Sodden, Jim explained that while they had an upstairs room in their home, he and his wife Emily were worried that the family, which included three children under 8, might get trapped up there with no means of escape. Shannon and Sean assured him that it was no bother, of course they could come over. After deciding to leave their dog with plenty of food and water in the upstairs room, Jim returned with a child on his shoulders, then went back for his wife and the two other children, who rode to relative safety on their parents' shoulders. Now, the twelve people sharing five hundred square feet had become seventeen, and the older children were joined by three much younger guests.

Emily was well aware that her family had exponentially complicated the situation when they arrived: "We felt bad—like, yes, they were letting us in—but we felt like we were imposing. 'Cause they were going through their own trauma. Like, everybody had their own trauma going on."

Her kids, though, seemed much better once they were inside and out of the rain. Now, they had other kids to play with. To Emily's chagrin, however, the younger boys discovered some intricate Lego creations Shannon's younger daughter had built and began to play with them: "Oh my God, the poor kid built these beautiful Legos . . . and my kids just destroyed them. And I felt terrible, even though Shannon was saying to her daughter, 'It's OK, it's fine, we have the instructions, we can rebuild it!' But [Shannon's daughter] was so upset. So upset."

When Shannon recounted this part of the story to me, she emphasized that she had purposefully guided the little boys to the Legos, so they would have something to do. Shannon's graciousness kept manifesting even after the Lego debacle. During their interviews, each of the other three mothers who were there repeatedly emphasized how kind and generous Shannon and Sean had been during this time. Both "hosts" continually waded into other parts of the house to gather food, water, and things for the kids to do: "I went and got [board] games out of the front. And they were really good about keeping themselves entertained and constantly eating on junk [food]."

It's likely that Shannon's focus on hosting helped keep her distracted during this period. Unlike some of the other mothers I interviewed who ended up with unexpected guests in their homes, Shannon genuinely did not seem to mind all of these people being in her flooded home, despite the general chaos. The ability to enact a role she was used to no doubt helped her retain a sense of control and normalcy as her home flooded, a pattern repeated across Bayou Oaks. Similarly, the role of guest—a guest who feels like a burden on her hosts—also repeated across the neighborhood. Leah sat on her roof for a couple of hours with her husband and two sons, ages 4 and 6, as the rain beat down. They sang songs, trying to distract the boys. Finally, she noticed a neighbor trying to get her attention from a nearby upper story. It was clear the neighbor wanted the family to come over and be safe inside her home. So they climbed down and waded over, joining several other people and pets already taking refuge on the second floor:

> So we hung out there and we still had power on our phones so we were texting, and messaging, and everyone letting everyone know where everyone is. And the kids were eating junk food, and playing games. And to them I think it was just like a really weird slumber party. Like that's how I describe it. Like... it was just kinda like, "We're at this lady's house. It's kind of weird and there's a lot of cats."

Back at Shannon and Scott's, after a few hours of the tight quarters, Emily felt like her kids were just too disruptive for the others. As she was sitting and processing what had just happened to her family's new home—she and Jim had just completed a six-figure renovation before they moved in not even two weeks before—her primary concern at that moment was that she was imposing on her neighbors. She could not abide the idea that her children might be complicating things for the three families who were already in the home when they arrived. Her mother lived nearby in a home that had not flooded, so Emily became obsessed with the idea of getting there. She wanted to stop imposing on her new neighbors just as much as she sought the comfort of familiar surroundings. It was difficult for Emily to just sit still and do nothing; she was used to being able to manage her family's situation on her own. As she explained, in these circumstances: "You just wanna be with people you know. You just wanna feel comforted. . . . [W]e wanted to actually make it to my Mom's."

She and Jim knew they needed to get to the neighborhood Kroger, where rumors suggested (inaccurately, as it turned out) that it was possible to be picked up in a car. Worst case, they reasoned, once they got to Kroger they could walk the next mile to her mother's home. It shows how desperate they were to get out of Shannon and Sean's hair that they even contemplated this journey. So in the still-pouring rain, they borrowed a pool float for their oldest, Melanie, put one son on each of their shoulders, and waded out to the front yard, where the brown, brackish water was over their waists. Emily managed a laugh as she described the scene to me: "So Melanie was lying back in this pool tube, with us holding onto the handles, and she says, 'This is the life! This is awesome! I love this!'"

As she and Jim stood out front, staring at the neck-deep rushing water in the street, they tried to decide if they should wait on a rescue boat or try to wade to Kroger. At that moment, Deborah, mother of a large brood, paddled by in her kayak. Known in the neighborhood to be a bit unusual, Deborah was an extreme example of hospitality norms during the flood. After water breached her first story, instead of retreating to her second floor, Deborah got her kayak out of the garage and started rescuing stranded neighbors and bringing them to her home, while her husband "ran a refugee camp" on the second story. All in all, Deborah ended up with thirty-eight people in her home that day. Mid-rescue effort, she convinced Emily and Jim to stay put. "Do you see that current?" she asked, pointing with her paddle. "It's not safe. You can't get across with your children." Five people were too many for her kayak, so Emily and Jim waved Deborah on and she paddled away. Just then, Melanie started shrieking, and Emily was horrified to see a floating mound of fire ants attacking her daughter. She and Jim managed to get the ants off of Melanie and agreed, exchanging a glance but not words, as the rain continued to beat down, that their escape attempt was finished for now. They went back to Shannon and Sean's, defeated. Emily was bereft, her sense of not wanting to be a burden so strong that looking back, she thought it was ridiculous: "It was so silly. Like hindsight is 20/20—you're just like, 'what were you thinking?' Really, we were thinking we don't wanna burden our neighbors."

The feeling of not wanting to burden their hosts was also impacting Ellen and Bill, who set out in their kayak to assess the situation at Kroger. Finding first responders and a crowd of neighborhood residents there, and believing it to be the best place to have a chance of getting a ride to somewhere dry, Bill left Ellen at Kroger and paddled back to Shannon and Sean's, returning with the dog and their son like someone in a logic puzzle brought to life. Finally, he paddled back once more and finally returned with their daughter. Reunited as a family, they handed over their kayak to some grateful first responders and boarded a National Guard truck heading to Houston's designated shelter, the George R. Brown Convention Center.

Meanwhile, as night fell, Shannon tried to make her guests as comfortable as possible, in hopes that they could get some sleep. While Sean heated up cans of soup on their camp stove, Shannon offered beer and wine to the adults. None of the adults really slept that night. The first floor addition was only four feet off the ground, and all of them worried—silently—that those four feet would not be enough. The local media were saying that it was going to keep raining for days. When would it stop? How high would the water rise? Without power, it was impossible to even see what the water level was outside. Was this going to be like Katrina? These thoughts kept the adults fretfully awake all night. The next morning, once the waters had receded somewhat, Emily and Laura and their families, still worried they were burdening Scott and Shannon, dispersed to other temporary accommodations. Before they left, Shannon pulled some pancakes she'd made before the storm out of the fridge, and everyone ate heartily. Sean even opened a bottle of champagne for the group. Emily got a little teary as she described Shannon and Sean's hospitality: "They're just the nicest people ever. Like, I don't know how you thank people for not only letting you stay there, but feeding you, like, the most amazing food." The nerves of the hosts and guests alike, though, were fraying, and enacting the norms of hospitality could only last for so long. As the second day dawned, some of the mothers grew determined to leave and began to devise strategies to get out. But where could they go? And would they be placing their families in more danger if they left?

At a certain point for each group stranded on an upper floor of a home, discussions arose about whether they should evacuate in one of the rescue boats going by or in a helicopter. The question sparked a lot of disagreement among the adults. Some argued that while they weren't comfortable, they were safe and had food, so they should just stay put until the water receded, and let the rescuers help people who were having medical emergencies or were really in danger. Others worried that since it was going to keep pouring for days, they might in fact *not* be safe where they were, and it was foolish to take the risk of staying. They remembered photos from Katrina of water up to the rooftops. Some were so uncomfortable without power or toilets that they just wanted to get somewhere dry and comfortable. Most of the families ultimately decided to evacuate the place they'd stayed during the storm. During this time of heightened danger and stress, many of the mothers engaged in a timeworn parenting strategy: portraying the unusual or scary as exciting to their children. Much as they had played up the element of family fun during the storm preparations, they remarked upon their experiences evacuating as an adventure for their children. This strategy served to reassure their children and also helped the mothers keep their own emotions under control.

Leah, who'd sat on her roof with her two young boys for hours before being taken in by a neighbor, eventually waded with her family out to drier ground and encountered volunteer rescuers in a high-wheeled pickup:

> LEAH: So then we get to like [an intersection] and there's this pickup truck. And . . . it's, like, helping people. So they're like, "Hop in." So we get in the bed of this truck, which my kids think is awesome. Like so awesome.
>
> INTERVIEWER: Had they ever done that before?
>
> LEAH: No. No, c'mon, these Jewish kids from Bayou Oaks?! In the back of a pickup truck!

Leah's kids had never before considered riding, or had the opportunity to ride, in the back of a pickup truck. So it felt illicit and exciting. Ellen also described marketing their evacuation experience as adventure to her kids: "They saw this really awesomely cool SWAT deep-water vehicle and, [I was] like, 'How cool was that!' You know? You know, 'You're gonna have something great to tell your kids and grandkids.'"

It was worth a try, the mothers thought, to frame the evacuation to relative safety as something exciting and memorable, but the mothers' careful control over the flood experience cracked further as the evacuations actually began. Julia's family waited out much of the storm in their playroom over the garage, but at night it was pitch black outside and in, and they couldn't see how high the water was getting, which was scary. Eventually she and Jeremy decided they needed to evacuate and hailed a passing rescue boat. The men with the boat told the family to take their time packing up what they might need, so Julia busied herself doing that while unbeknownst to her the men floated her children out to the boat on an air mattress:

So at this point I turn around and I said, "Where's my children?" He's like, "They're on the boat, they're safe, ma'am, don't worry." And I'm thinking "Don't worry? I don't see my kids. You're gonna see somebody else who needs to be rescued, [if you] move with that boat." I am beyond . . . ."worried" doesn't cut it, okay? I am 150% not okay with the situation right now. "Get me to my kids now!" And I flipped. Because, I mean, I'm a psycho in a *good* situation.

Julia was having a hard time not being in control of the situation that she'd planned for so carefully. Someone else was now calling the shots, and this evacuation was not part of her plan. But she hadn't lost her sense of humor, as was evident as she continued her story, describing how she and Jeremy packed for the evacuation: "I'm leaving like I'm checking into the Marriott. I'm going through water, but . . . we had suitcases. We had duffel bags. We had backpacks. We had 16 different kinds of technology. We brought, you know, the dog in his crate. I brought the [kids'] loveys."

It was hard to shake the trappings of packing for a typical trip the family might take. After all, it was unclear how long they would be out of the house. Once her family and their myriad belongings were safely in the boat, Julia breathed a little easier. She attempted again to conjure the adventure mirage, but wasn't sure her older children were buying it by this time:

So now it became, "Guys this is gonna be an adventure. We're gonna get on a boat." And I'm trying, but I mean they're not stupid kids. They know mommy's stressed out. They know we don't know . . . we don't know what we're doing. Though we kept saying, "Have you guys ever been on a boat before through our neighborhood? It's so crazy . . . isn't this so crazy? Oh my gosh this is really nuts." Like we tried to laugh a little bit, make it an adventure. We kept saying the word adventure, and trying to make it like that.

Julia's efforts to mask her stress were substantial, but her kids could see through them. As the evacuations proceeded throughout the neighborhood, that sense of adventure and excitement became harder and harder to sustain.

As parents and neighbors deliberated about evacuating or staying put, and word spread that people were being taken to Houston's George R. Brown Convention Center (which Houstonians call the "GRB"), racialized images of the Superdome during Katrina loomed large in the mothers' minds. Although the worst stories later turned out to be false or at least overblown, this was not widely known, and memories of the crowded, dirty Dome and its mostly Black, mostly poor, temporary inhabitants were still fresh more than a decade later.[4] While I expected the mothers to say that they rejected going to the shelter because they were not in dire enough circumstances, instead, they clearly rejected the idea on the basis of their perception that it would not be safe—a perception clearly shaped by both class and race. Many of them referenced "danger" and "chaos" when I asked why they did not want to go to GRB. Rather than seeing it as a place of refuge, the mothers largely considered GRB a potentially dangerous place for their families, one to be avoided at all costs. Their own neighborhood, a carefully chosen bastion of safety in normal times, which had now been transformed into objectively one of the most dangerous parts of the city, was where they wanted to stay. They were willing to bear the excruciating uncertainty of staying in a flooded neighborhood, with yet more rain forecast to fall, rather than be subjected to a situation in which they could not control who was around them or what would happen next. Their desire and ability to curate the flood experience, they felt, would be completely undermined, potentially exposing their families to trauma. As Cathy put it, summarizing feelings I heard from many mothers:

> Then [the adults] all kinda took a vote, like Lord of the Flies, it was really tense. 'Cause we were like, "Do we wanna go to a shelter? Is that where we wanna go? Or is that gonna be more traumatic?" And I thought of like horror stories from Katrina that you heard about the shelters, and I'm thinking like, sexual assault and all kinds of things.

Leah, like many of the mothers, explained it this way:

And so like the whole thing that day was like people were like, "Well, do you wanna be rescued?" And I was like "Well, where am I being rescued to?" Because right now my kids are at a weird slumber party where we have food, we know we're safe, and we're all together. I'm pretty sure George R. Brown will traumatize me. Some of it is the whole idea of, like, remembering the Katrina Superdome, and obviously this went a lot more smoothly, but how are we to know that [at the time]? But the whole idea of just like, masses of people, and chaos, and you know masses of people and chaos. And like, you know, the dangerousness of maybe too many people with nowhere to go, kind of thing.

Both Cathy and Leah, along with many of the other mothers, emphasized that in hindsight, they were grateful they stayed put and did not evacuate. They felt they had an easier and less traumatic experience than the families who did evacuate and go to GRB. A shelter was just not for them. While they certainly saw themselves as impacted by a disaster, they did not feel they were like the disaster victims splashed across the news after Katrina. Those were the people, poor, Black, and Brown, who needed a shelter, not them. Some mothers felt so strongly about not going to GRB that they turned away first responders and rescue boats trying to help them. Leah, who at the time was literally sitting on her rooftop with her family as the rain poured down, described how she was not interested in a rescue:

> We saw all these helicopter rescues. And we're kinda like more making it like [to the kids], "Look, how cool." Like not "Oh my God." It was more like, "Look at all the helicopters," and you know they'd be like waving... we'd be like, "Don't wave! We don't want them to come here."

Amy's older daughter was disappointed her family wasn't going to get on the boats she saw going by outside, because she thought it would be fun. Amy, however, was not interested in a boat ride:

> She had no idea where it was going or what that meant. And in retrospect I'm like, we would have ended up at George R. Brown. 'Cause that's where... it was an official boat and that's where I now know other people ended up, so I'm glad we didn't get on it.

Mary and her family, who along with several neighbors actually broke into an empty two-story home behind their house to escape the floodwaters, also rejected a rescue boat: "I said, 'I'm not goin' to George R. Brown,' 'cause I already knew that it would be disorganized, and chaotic. So I did not want to go to George R. Brown. That wasn't an option for us."

Not all families managed to avoid GRB, however. Most of those evacuated by boat ended up there after a harrowing ride in the back of a National Guard vehicle. But even if they ended up in a National Guard truck, they didn't necessarily go into the shelter. Instead of seeing the shelter as a place to rest and recover, several mothers saw the trip to GRB as instead a way to get a safe ride to downtown, which was accessible by car from a few routes. Two mothers stepped off a National Guard vehicle and then just walked away instead of checking into the shelter. One of those families was picked up by a friend, and the other simply walked to a nearby hotel and checked in. Jennifer and her husband and two boys checked into GRB and stayed for a few hours, but ultimately decided to leave because they didn't feel safe. They tried two nearby hotels, which were both full, before deciding to try the Four Seasons. As Jennifer put it:

> Fortunately they had a room available even though it was very expensive. But at that point, after all that experience, you know being flooded . . . being rescued by boat . . . and the whole time it was raining on us. And we're just cold and so tired. And we said, "I don't care, we're just gonna stay in the Four Seasons," you know . . . because it will just be a one-day or two-day deal. But it ended up we stayed like four days . . . or five days. I can't remember. Yeah. But you know still, we did not regret it.

Only two mothers out of thirty-six ended up actually spending the night at GRB: Julia, who'd evacuated with her family, the family dog, and many suitcases; and Ellen, who'd kayaked to Kroger in shifts with her husband, their twins, and their dog after leaving Shannon and Sean's.

Julia and her family spent what felt like an eternity waiting in the Kroger parking lot, as rain beat down. "Everyone looks like a refugee. At the same time, you're finding people you know . . . it's weird. It's like an upper-middle-class refugee situation." Jeremy managed to scrounge up some garbage bags, which they wrapped around their bags and suitcases in an attempt to keep them dry. Emergency personnel were loading people onto dump trucks, but Julia didn't know who they were or who they were working for. She also didn't know where the trucks were going, but the rumor seemed to be the GRB. Julia's discomfort was similar to that expressed by other mothers: "That's an unsettling situation. I don't know anything about what's going to be at GRB. I don't know how to get out of GRB. It's far. Is this, like, a Katrina situation? I don't want to be in a Katrina situation."

Eventually, it became clear that unless they wanted to continue to sit in the rain in the Kroger parking lot, they were going to GRB. Julia's family was loaded into the dump truck last, directly by the wide-open back. The emergency personnel sternly warned everyone to hold on to the sides so they wouldn't fall out of the back. She told her son to hold tight to her middle daughter. The youngest girl was struggling, but she wasn't being difficult—she was just quietly terrified. Julia huddled over her, trying to shelter her from the rain, soaked to the bone and shivering, despite the August heat. Close to midnight, the truck finally arrived at the shelter downtown, having traveled a roundabout route for over an hour, a tidal wave of water sweeping over its occupants every time it turned a corner.

Once finally inside the shelter, soaked to the bone and freezing in the intense air conditioning, Julia gratefully accepted blankets for everyone, broke down in tears, and hugged the volunteer who was giving them out. Then the family submitted to a search of all their (numerous) bags and a pat down, to ensure they did not have weapons. Julia was disconcerted by this experience, which she considered bizarre. "Like, I'm really glad. Thank you. But would that have ever occurred to anybody? Never. Like, oh my god, they're searching me." Julia, like the other mothers, was not accustomed to situations in which she was treated as a potential criminal. Eventually someone pointed them in the right direction, and Jeremy set up camp with their belongings and some cots while Julia inspected their suitcases to see if anything was dry. Miraculously, the kids had dry clothes to change into, but everything in Julia's suitcase was soaking wet. Luckily, there was a large pile of donated clothes in the center of the room which Julia sifted through for something that might fit her. In the restroom, a mass of wet and frustrated women were trying to clean themselves up too. Her kids, by this point, were complaining, and Julia felt annoyed and embarrassed by their complaints:

> I mean you want them to shut the fuck up because you know some of the things they're complaining about are like, "That's great that you're hungry, kid. You know what? That guy hasn't eaten in a month." You know? "Mom there's no soap!" Obviously. You know? I'm like, we're in a little bit more desperate situation than they seem to realize.

Julia and the kids hastily changed into dry clothes, but with no other choice, Julia had to leave on her sodden underclothes. The family then spent some time trying to figure out what was going on—ascertaining the rules and rhythms of this strange environment. When they saw people coming back into the room with food, Jeremy went out in search of it—and brought back a couple of sandwiches and granola bars. The family pushed their cots together, ringing their suitcases to protect their belongings, and tried to sleep. As it turned out, however, the miserable dogs in the cavernous room barked all night long, so no one got much sleep. When I asked Julia if she'd ever felt unsafe, she was emphatic that, contrary to the worries of the other Bayou Oaks mothers, while she'd certainly felt out of place, she did not ever feel like their safety was in jeopardy, because everyone just kept to themselves: "Everybody was very . . . everybody was dealin' with their own shit. There wasn't a sense of, 'Let's all help each other' even. Everyone had their own shit to deal with. I never felt unsafe."

Even though initially Julia had been uncertain about going to GRB, once there she realized that it was not an experience to fear. In the morning, Julia felt ridiculous as she shepherded her kids through the typical routine of brushing their teeth in the packed, smelly restroom. Her mothering routines were so ingrained that they persisted even in this unusual situation:

> And so in the morning we get up and I brought the kids to go brush their teeth in the bathroom 'cause that's what you do when you wake up. You go brush your teeth. By the morning, oh my god . . . that bathroom deteriorated quite a bit in those few hours. And it smelled very, very bad. And there was water, and stuff, and random articles of clothing everywhere. And like a conga line of women fighting over the plug, 'cause there are no plugs in the whole [shelter] . . . literally. So yeah. And here I am parading my eight-year-old saying, "Go brush your teeth dear!"

Ellen's dump truck ride to GRB with her family was similarly harrowing. Their truck stopped for more than half an hour on the freeway in the pouring rain. An older woman on the truck sobbed loudly the entire time. Ellen's family huddled up, with a poncho over their heads. Ellen, who had had her emotional moment at home, wasn't emotional at this time. She was just frustrated to be sitting on the freeway in the rain, with no one seeming to know what was going on. Their dog was traumatized, as Ellen put it, and Bill was fretting that someone was going to get fed up with its noise and kill it. Eventually, the truck started going again and they got to GRB, more than an hour and a half after they were first loaded into the truck.

Once the family arrived and got checked in, they were assigned a room and waited in a long line to receive a cot. While they waited, Ellen was taking in her surroundings:

> And so they were literally putting them together and you know pallets were coming in and volunteers were putting them together as they were coming in. And we saw some neighbors there. But, you know, I have to say the vast majority of the individuals who were there were not necessarily our contemporaries. And so it was a little disconcerting so to speak. But that being said, it was you know what it was. Where else were we gonna go?

Like Julia, Ellen clearly felt out of place during her GRB experience. Similar to the vast majority of the mothers who purposefully avoided going to GRB, they felt like the other people in the shelter were not their "contemporaries." Unlike in their carefully selected neighborhood and school, at GRB they could not control who was around them or what their children saw or heard. Both Julia and Ellen, however, reported that while initially disconcerting, their experience at GRB was not actually traumatic. It was chaotic at times, and their families shared space with other flood refugees from all backgrounds, but it was in fact a safe place to sleep, find some dry clothes, and have a meal.

Although researchers and journalists have debunked many of the myths about what went on at the New Orleans Superdome in 2005, images and stories from the media dominated the thinking of these families during their own time of crisis. While the mothers, as a group, expressed progressive ideals about race and class exposure for their children, like other upper-middle-class and affluent parents, they also wanted to curate those exposures.[5] Being thrown together in a shelter was not the type of exposure they imagined their children (or themselves) having, and they actively avoided it. It is important for disaster researchers, first responders, and policy makers to understand how perceptions of the post-Katrina experience may shape evacuation behavior in other disasters. As it turned out, for the Bayou Oaks families, staying put in upper stories of homes was a safe decision. But they had no way of knowing that when they made the choice to refuse assistance from first responders and rescue boats, a decision that was clearly shaped by both class and race.

To truly grasp Houston's Harvey experience, it's important to understand that this was not a hurricane that blew through in a few hours. This was a massive, slow-moving megastorm whose rain bands stretched nearly three hundred miles. It rained for days. Also, unless you've lived in a tropical climate, you can't understand what Houstonians mean by "rain." Houston rain is not gentle or misty. It pelts down with drops the size of dimes, and in two seconds, you're soaked to the bone. Your hair is drenched. Your shoes, ruined. And that's just a regular summer rainstorm. Harvey brought torrential, tropical downpours. Floodwaters did not recede for days or weeks in some areas. If your neighborhood escaped flooding from the first few bands of rain, you were still in danger from those that followed. Many neighborhoods west of the city, in the far suburbs, were flooded when reservoirs overflowed, days after the storm. The entire city didn't exhale for nearly a week. So while most of the mothers and their families believed they'd gotten to safety within the first forty-eight hours, sometimes that feeling of safety did not last.

Rebecca, whose home had flooded on Memorial Day 2015 as well, evacuated to the hotel on the west side of town where her husband Paul worked, ahead of Harvey's landfall. She had no interest in her family experiencing another flood and was happy to get out of Bayou Oaks and onto the fourth floor of a solid-feeling hotel while the hurricane lashed the city. After a couple of days of eating in the hotel restaurants and trying to monitor remotely what was happening back in Bayou Oaks, the family got word that the hotel was starting to flood and they would have to leave. Rebecca raced around to get the essentials packed up into just a couple of bags, tucking her handle of vodka into one. "Is this a necessity?" asked her 11-year-old sarcastically. "If I end up in a shelter, it sure is," Rebecca replied.

After seeing on the news that her hotel was being evacuated, nearby friends got in touch with Rebecca and offered to come and pick up the family, as routes west of the hotel were still passable. That night, Rebecca was getting her kids ready for bed in their second temporary quarters of the day. She allowed herself to exhale for what felt like the first time all day. Paul was reading out loud from an iPad while cuddling their 3-year-old. Her older son started to fret about going through flood recovery—again: "He said, 'What are we gonna do about' something? And I said, 'I don't know. I said, 'I can only kind of focus on one thing.' And he's like, 'What?' And I pointed and I counted us off and I said, 'One, two, three, four . . . we're all here.'"

Once the mothers were safe, either with friends or family or in some temporary lodging, they could start processing what had happened to their families. They could finally begin to regain control over the experience. Although some minor injuries were sustained during the storm and its immediate aftermath, all of the families emerged intact. The major work of recovery—the calls to insurance, FEMA, contractors, and adjusters; the finding of someplace to live; cleaning out their ruined homes—could wait until tomorrow. For the moment, the mothers were grateful to be safe and that their families were together.

4 Storm Recovery

YOU CAN FEEL SORRY FOR YOURSELF WHEN THE WORK'S DONE

Once the storm had subsided and the waters began to recede, the mothers were anxious to begin cleaning out their homes. They wanted to see the damage and know what they were dealing with. They also knew that the longer the sodden furniture, household items, and drywall steamed in the brutally hot and humid houses, the worse the situation would get. They also had to begin the time-consuming and heartrending task of documenting their losses for insurance purposes. A mountain of tasks lay before the mothers, but they were eager to tackle them because everything they could accomplish quickly would get them that much closer to their end goal: getting back into their family homes. For the mothers, their homes were the center of the status and stability they curated for their families. Nothing could be made right, not completely, until they were back in those homes.

Throughout the recovery process, the mothers' goal to get back in their homes quickly was supported by their access to financial and social resources enabled by their status as upper-middle-class families. Although these benefits would continue to accrue across the recovery year, they were concentrated primarily in the first few weeks after the storm. These benefits took many forms: the mothers had access to liquid assets, both their own and those of family members and friends; they had extensive social ties through alumni organizations, workplaces, religious institutions, and community groups; and those ties included plenty of people who had access to time and money. Many of the families, and in particular the mothers, were able to convert these social connections and financial assets into a streamlined recovery process. This chapter explores the depths of resources, both social and financial, the Bayou Oaks families were able to draw upon to restore their homes and neighborhood. Through their stories, we can see not only how disasters might increase inequality in communities, but also how for all these important advantages deriving from wealth and privilege—combined with the mothers' organizational skills and converted into "class-related efficiency"—the families nonetheless still faced an uphill, expensive, messy struggle to get their homes cleaned out and their renovations started.

Going back to their houses for the first time after the storm was a gut-wrenching experience, even for those mothers who had been flooded before. They vividly described scenes of overturned furniture, collapsed bookshelves, felled refrigerators and spilled contents, and household items that had journeyed far from where they'd started. Several mothers described the scene as "like a war zone" as they attempted to convey the destruction they saw and the devastation they felt. For many, all the items they'd carefully stacked on top of kitchen counters, on dining tables, and on beds before the storm had been submerged by the water, which rose much higher than anticipated. So when the water finally receded, leaving a stinking, filthy grime on everything it touched, the homes' interiors were unrecognizable: "It was like everything had been put in a blender, shooken' up, and then sprayed everywhere."

Many of the families assumed, during their preparations, that items placed on top of heavy furniture pieces like entertainment centers and wardrobes would survive as the water flowed around the furniture. Not so, as Meghan, a twice-flooded mother of two teens, remembered:

> You know it was crazy. The furniture was all upside down. I mean [my husband] had warned me [before he evacuated] how high the water was 'cause it was almost to the kitchen counter. And you know, we have these two huge credenzas that were like solid wood. It took like five guys to move them in. Floated and flipped upside down. Like how does that happen? You know?

Bookshelves collapsed, sending special books and wedding albums stored high up and supposedly safe into the water. Even sturdy dining tables collapsed or flipped over, spilling family heirlooms and china into the abyss. Mothers also compared the scenes to what it looks like—and feels like—after you've been robbed: "It feels like someone has been inside your home and gone through all your stuff. And you just felt violated."

And the families *had* been robbed in a sense: robbed of their sense of stability and calm, along with most of their possessions, even those they'd thought to be safely stowed out of danger from the water. In addition to the terrible sights, when the families came back home, they were greeted by searing, swampy August heat and unbearable smells. Food had begun to spoil, sending the stench of rot into the air along with the fetid stink of the waterlogged belongings. Mothers described the smell as "like sewage topped with gasoline," making them gag and choke. Their homes, which they had worked so hard to cultivate, were not only destroyed, they now felt and smelled like toxic waste dumps. The sense of safety and stability they saw as their job to provide for their families was gone, and now their number one priority was to get it back.

Basic knowledge of how to remediate a flooded home was widespread in Bayou Oaks before Harvey. Flood capital—this time, for recovery procedures—was again an important currency throughout the neighborhood, shared via social media and between neighbors.

Flood remediation was a five-step process. The first step was to turn on the air conditioner and set it to the coldest temperature possible to inhibit mold and start drying out the house. Next, families needed to document the loss with hundreds of photographs, critical to maximizing a flood insurance claim. They also had to create detailed "contents spreadsheets" to document their losses, including when and where items were purchased, their original cost, and current value. Third, items had to be sorted into those that must be discarded and those that could be cleaned. Fourth, saved items had to be cleaned. China and other valuable kitchen items were often washed, soaked in a bleach solution, and then washed again. Clothes, even those that had not touched the water, needed to be washed in a special solution or dry-cleaned, because the stench permeated fabrics. And finally, once the house was clear of belongings, the house itself had to be remediated. Wet, heavy carpets and rugs had to be ripped up and hauled to the curb. Drywall needed to be cut out well above the water line, and hardwood floors pried up and discarded. It was heavy, hot, dirty work—but luckily for the mothers of Bayou Oaks, help was on the way.

Ashley, the steely, blond, 38-year-old principal of Bayou Oaks Elementary, was assessing the extent of the damage at the school building. She received word from the district that personnel were barred from visiting the school grounds for safety reasons, but there was absolutely no way she was going to follow that instruction. As soon as she heard it was possible to get to the neighborhood without flooding a vehicle, Ashley was off like a rocket. She went several times over that first week post-Harvey and was repeatedly fussed at by district employees. "It was weird—it really seemed like they had me chipped. They always knew when I was there!" She didn't care. This was her building. It was her community. She needed to assess the damage and see it for herself. The first time she went, she sloshed toward the front doors through the inch or two of standing water that still covered the site. She immediately noticed the water line on the outside of the building; it was at least three feet high. Ominously high. Ashley was greeted by the manager for the remediation company sent by the district. "Hey," he said, "are you the principal?" "Yeah," Ashley said. The guy grinned. "Well, you won! You have the most damage in the district!" Ashley took a moment to absorb this. "How bad is it?" she asked. He quickly grew grim. "It's pretty bad," he said. "We'll get it cleaned up, but I don't know that you'll be back in this building." This was the first time it occurred to Ashley that the whole school building might be gone for good.

The workers would let Ashley inside, but not her two staff deputies. While she walked through the building and viewed the extent of the damage, she had a sinking feeling. It was even worse than she'd imagined. Crews of men were swarming the building, discarding, packing, and cleaning. The musky, sewage-laced smell inside was overpowering. The classrooms looked as though they had been ransacked, then hit by a dirty tidal wave. In that moment, Ashley knew she had to let go. She could not save this building. She could not ever bring children back into this toxic place. Nothing could clean it properly; nothing could make it safe. True to form, Ashley made a quick executive decision and refocused her mental energy. She could see there was nothing she could do at the school, but there was plenty she could do outside it. When she finished her tour, she said nothing to her staff other than that it was bad inside. She felt strongly that she needed to tightly control the information, so that panicky rumors would not start to spread. "I didn't want the school business out. I was so protective of the school." She could not even tell her top lieutenants. She had a deep understanding of her school community and how it operated, and just how quickly misinformation could spread along the dense web of social connections. The last thing she wanted was for her flooded families to worry about the school right now. She knew they would fret about the loss of the place that meant so much to them and panic about the school community being split up and sent to different schools. There would be time to worry about that later.

Ashley knew firsthand how her school families were feeling. In 2005, her parents had lost their dream home, a rambling, "'John Grisham, iced tea on the front porch' kind of house," on the Mississippi Gulf Coast, to Hurricane Katrina. Hours before the home was destroyed by a direct hit from a tornado, Ashley, who was then working as a teacher in Mississippi, enjoyed a family dinner at the kitchen table. Eying the radar, she managed to convince her family to evacuate to her apartment a few miles north of the coast. When they returned to the house the next day, all that was left was a scrap of that table. Ashley knew what it was to lose everything. But to her knowledge, no one in the Bayou Oaks school community knew her story. "It's just not about me," she said, when I asked her why not. But Ashley knew what would be needed and how to make it happen. She knew that the families would want to be back in their homes as quickly as possible, and that extra labor would enable that. "Here's what I need you guys to do," she said to her deputies. She instructed one of them to be the point person for volunteers and the other to be the point person for people who needed help. Although her two staff members were new at their jobs, Ashley did not have time for hand-holding. She remembered that she was very stern when she spoke to them: "They call it my 'principal voice.' It was go-time."

Glad to have a task to manage, the women set to work and staged a command center in the front yard of a school family's home. They coordinated supply drop-offs for items like buckets, cleaning supplies, garbage bags, and plastic bins. They spread the word that the command center was open for business. Offers of assistance for the school, like book and supply drives, were pouring in but were not helpful at that moment, because there was no place to store anything. (Ashley recounted her attitude: "Don't send a Single. Freaking. Crayon. Here!") All of these offers, nonetheless, needed to be responded to and tracked. Families who had been flooded needed to be enumerated and visited. Teams of volunteers needed to be formed and dispatched to homes. The command center also tracked specific needs of families, like mattresses, children's clothes and shoes, loaner vehicles, meals, and medical supplies. Ashley sent a message to all the teachers that while there was no obligation to participate, starting the next morning at 8:00 a.m. volunteer teams would help families clean out homes, and she anticipated seeing many of them there. But Ashley had no idea how many, if any, would come to help.

She needn't have worried. The teachers showed up in droves across the ten days of the effort and fanned out across the neighborhood, ripping out drywall and packing up or discarding household items. Mothers greeted them like old friends, crying and hugging them. Just as the Bayou Oaks mothers deeply valued the school community, the teachers and school staff valued them. One teacher in particular, the taciturn pre-K teacher Ms. Jones, who'd been at the school for more than thirty years and inspired affection, respect, and yes, more than a little fear in the parents, was mentioned repeatedly as a calming influence. As Janet, mother of two, put it:

> Well, we had a lot of teachers. People from [the school] . . . you know they had a group sent over. Ms. Jones is fabulous. She is amazing at getting things done, and telling the other teachers what to do, and what not to do. I was so happy when she was there. I mean all the other teachers were great, but you know, it's like Ms. Jones is . . . it's just Ms. Jones.

As Cathy, mother of three, put it: "Like, Ms. Jones would be like, 'listen up people, this is what's happening!'" When I mentioned to Ashley that Ms. Jones was coming up a lot in my interviews, she laughed. "Oh yes—well, she's the matriarch." It was clear that the mothers took great comfort in the school staff's efforts. They felt cared for: not just their children, but the entire family.

Bayou Oaks Elementary was a critical institutional support for the neighborhood. It served as the community's emotional center, and its strong network, headed by a dynamic leader, sprang into action after the storm. This network was also largely female. Although there were some school fathers who were active volunteers, the mothers were the family links to the school, the people who attended most of the Parent-Teacher Organization (PTO) meetings, talked with other mothers out front while waiting to pick up their children, read all of the emails and social media posts, and signed their husbands up for occasional volunteer jobs. Ashley and her two deputies were also women. All but one of the school's teachers were female. All of this labor—the organization, the distribution of supplies, the management of volunteer teams, and the physical labor inside the homes—was achieved without support from first responders. It was not paid for by FEMA. It was not reported on the local news. The mayor did not visit. No one was recording the hours. These were private efforts, organized by a public school—the key institution within the neighborhood—and while men showed up to help with the physical labor, the organizational labor was largely female. Such "hidden" but crucial contributions of women to disaster recovery have received little attention in the literature, with only a few exceptions.[1] As sociologist Elaine Enarson wrote, "When we cannot see or appreciate the significance of 'what women do' in disasters, we cannot capitalize on the skills, resources, and local knowledge of women and women's community-based organizations."[2] Bayou Oaks Elementary women—the principal, her staff, her teachers, and nonflooded mothers—dropped everything and focused their considerable talents on the matter at hand: restoring their community. Just as women's emotional and cognitive labor within households is often hidden and underappreciated, so too is their social support and volunteering labor.[3] The mothers' groundwork of building social connections before the storm, something they viewed as part of their family responsibilities, was primarily focused on the curation of the family life they desired, but the flood activated and exposed the deep network of women who tended to the neighborhood and its residents.[4]

It was not just the school community who showed up to help. Some of the mothers reported that at first they worried about how they would ever get their homes cleaned out, because the damage was so much more widespread this time, and most of their neighborhood-based social circle also had been flooded, so they were busy with their own homes. But quickly they realized that people they knew only slightly—acquaintances and friends of friends (what sociologist Mark Granovetter called "weak ties")—were showing up in droves.[5] While many neighborhoods, and not just affluent, White ones, received help in Houston during the week after Harvey, the help in Bayou Oaks was over the top. Locals knew the area was hard hit, and not for the first time. In addition, the Bayou Oaks families had connections to many different local institutions and organizations, which in some instances they purposely activated and in other instances were activated on their behalf. Alumni groups from schools and colleges, employer-sponsored volunteer crews, Boy Scouts, local sports teams, and work crews organized by synagogues and churches descended on the neighborhood. While many of these groups targeted homes where they had connections to the residents, some just went door to door until they found someone who needed help. The Jewish Community Center, which had also taken on more than four feet of water, nonetheless was soon operating a drive-thru distribution center in the parking lot for free cleanup supplies.

Practically as soon as the mothers arrived back at their homes, friends and family members began calling to see if they could help, or just showed up at the house, ready to work. The volunteer teams coordinated by Bayou Oaks Elementary, mostly teachers, were dispatched to school families' houses. And roving bands of volunteers—sometimes connected to a family by a weak thread, sometimes strangers—showed up. At this point, most of the mothers immediately snapped into their familiar household manager role, quickly deciding what needed to be done and in what order. Accustomed to commanding teams at the office or marshaling small armies of PTO volunteers, the mothers applied their organizational and executive skills to the matter at hand.[6] While during the flood and evacuation process itself, the mothers had begrudgingly relinquished control, the aftermath was something they could manage. As Julia put it:

"We go into that [managing] mode 'cause now this is where actually I think I'm really, really good at this. So let's go!"

Now that the waters had receded and her family was out of the shelter, Julia was in her element. She knew how to manage a moving and clean-out process, having helped other friends after the prior floods. She was back in control and knew what to do. This was her sweet spot. For Kelly, 48, mother of two, the management role was literally what she did every day at her job in the medical industry:

> And then I go into like planning mode, 'cause I'm a project manager. And like, okay, we've gotta get the electricity off. We've gotta get whatever. Okay how are we gonna get drywall. How are we gonna get. . . . You know we've gotta get moving. And so then the work began, right? Like as soon as the water went down.

Meghan, having secured her rental house by wiring a year's worth of rent as the rain was still falling, had a key advantage that most of the others did not: a clean, dry place ready for people to take household items to. She was therefore several steps ahead of even the most organized Bayou Oaks families, having even had the foresight to print out directions for volunteers to the rental home that she could hand out. And she also had a plan for keeping things organized and reducing her own packing and unpacking time:

> So I had these directions [to the rental house] and I printed out several copies and said, "This is what I want you guys to do. Like pick a car, that will be [daughter's] room. Take all the shit out of [that] room, put it in one car, so when you get over here everything in that car goes into her [new] room, right?"

Rebecca, mother of two and also twice flooded, similarly sprang into action, her prior experience having taught her about the importance of speed after a flood:

> I immediately just started. A lot of people like to flail their hands for a while, and feel sorry for themselves, which I know some people emotionally need to do that. But I think it's a waste of time. I think

you can feel sorry for yourself when the work's done. Because . . . in a flood it's not like homework where it's the same amount of homework if you do it today or if you do it tomorrow. [With] the flood, it gets to be more work the longer you wait. So the faster you get in there, the drier things are. It's just easier work to do immediately.

As Rebecca points out, a key piece of flood recovery is the efficiency with which it is carried out. The longer you wait to do the work, the grosser the house gets and the harder it is to take care of. The mothers' flood capital from prior flood events taught them this lesson, as well as what they needed to do to take care of it.

While about three-quarters of the mothers, like Julia, Kelly, Meghan, and Rebecca, were immediately on top of managing cleanup, the remaining quarter, almost all first-time flooders, were paralyzed by the amount of work that needed doing and first needed time to emotionally process the disaster. Some didn't even know where to start. These were families who didn't expect to be flooded, and they hadn't prepared as thoroughly as the others. Sarah, nursing a preflood injury, was frustrated because normally she would have just tackled the job entirely on her own; she saw efficiency and organization as part of her identity as a mother:

> And then that first day I just remember going in and . . . just kind of going around in circles. You don't know what to do. You just look around at everything and you're like, "Where do I start?" You just didn't know where to begin. Because, it's my personality—I would have packed up the whole house by myself [if I weren't injured]. You know what I mean. That's just a mom thing. But I couldn't do anything, which was equally frustrating. And then that's when this community just kicked into like a gear I've never seen before.

For Sarah, not being able to jump in and manage the process was difficult because she saw it as part of her role as a mother. She was not used to "going around in circles" while she decided what to do.

Emily, who'd moved into Bayou Oaks only twelve days before Harvey and took refuge with Shannon and Sean during the storm, also found herself at a loss when she and her husband went back to their flooded house. As she was dithering around in the house, a work friend contacted her:

> So it was really funny, 'cause she texted me and asked how I was doing. And I was like, "I am scooping water off the ground with a cup." And she . . . she didn't even respond. And I was like, "Oh that's weird." Twenty minutes later she knocks on the door. She'd shown up with mops, and buckets, and her friend who lives [in Bayou Oaks] had flooded Memorial Day. So she had gone and helped her [then]. And so she kind of knew what to do. And I was like, "Thank you, you didn't have to come." And she was like, "This is pitiful . . . what you're doing." She was like, "I couldn't believe it. I brought you mops."

For Emily to snap out of it, she needed her friend to show up and tell her she was being "pitiful" and show her what to do.

Maria, a financial professional, had been out of state during Harvey on a family vacation. She and her husband were not able to return to Houston until more than a week after the storm. When they got home, they found that everything had just been sitting and stewing in the heat, untouched, stinking beyond belief. Almost immediately, however, Maria's best girlfriends showed up and helped to lighten the mood:

> We had only been home for maybe an hour when everybody showed up with husbands, and kids, and hammers. I mean we got all the drywall knocked out in about three and a half hours. [A friend] had gone back [to her house] to check on the kids and came back with some water, and some snacks, and a bottle of champagne. I mean it was just . . . that is the spirit. We will show up and we will do the dirty work. But we're gonna do it drinking champagne 'cause that's how we roll.

In Maria's friendship circle, it was important to show up—and to show up prepared, with champagne.

In fact, the mothers reported that some of the most extraordinary and surprising help received during this initial phase was not from close friends or family, but rather from "some mom [they'd] met once at a PTO function," or a sorority sister they'd only kept up with since college on Facebook. When Nicole, mother of two and a first-time flooder, showed me her cozy, donated temporary apartment in a hip neighborhood near downtown, she gestured to two twin beds and said she wasn't even sure where those had come from, but that someone had brought them for the family's use. Jill, a two-time flooder, had friends who helped her find a rental home and then, while Jill was still cleaning out her flooded home, furnished the rental with borrowed and donated furniture and household goods and even stocked the refrigerator. When Jill finally left her flooded home and went to see her temporary digs, she was amazed to find a functional space, along with a retired Bayou Oaks Elementary teacher who was cheerfully putting new sheets on the beds while someone the teacher had hired was mowing the grass outside. Anna, who worked in the medical industry, received an unexpected text message one afternoon during the cleanup. One of her colleagues, a surgeon, had called his secretary on speaker phone during an operation to have the secretary check on Anna. "'What does she need[?] ... I have an extra car and a king-size bed. Does she need it?' From his operating room. The man's operating. So knowing that I have a work support system is phenomenal."

Importantly, many of the people in the mothers' social networks were of the same upper-middle-class backgrounds as the mothers. So the Bayou Oaks mothers were tied to a deep bench of people who could tangibly help after the disaster, and who were connected to *even more* people with the ability to help. As Allison, a twice-flooded mother of two, put it:

> We had my cousins show up, who live in the area. And another friend showed up who has sons at [a private high school]. Some people showed up from my husband's work. And a family from Little League, who also is a [school] family. We don't know them well. They just showed up. So just socially ... we were connected to everybody, but no one knew each other.

Allison's statement illustrates how the social networks operated. The family had connections to multiple organizations, and the people within those organizations had other connections. Everyone was willing to help immediately after Harvey, even friends of friends. As Allison put it, "We were connected to everybody, but no one knew each other." These connections were extremely valuable when the families needed them most.

It wasn't just locals who helped the mothers out. Friends and family also came up with creative ways to help from afar. Amy's friends, scattered across the country, organized to complete the spreadsheets that she would need for her insurance payout, based on the detailed photos she sent them of her destroyed belongings. This saved Amy an enormous amount of time and energy. Other mothers had out-of-town friends who called local apartment complexes, compiling lists of available units, amenities, and rents, sending them comprehensive emails with the information.

This massive outpouring of help, of course, is usually not the case for poorer disaster victims.[7] While the poor also have social networks, they tend to be smaller and less diverse than the networks of wealthier people and to involve more intensive and potentially short-term, "brittle and fleeting" relationships.[8] Moreover, the networks of lower-income people typically do not have the financial or time resources to provide help on the large scale that is needed after a disaster.[9] In the case of Katrina, class impacted how networks could be activated: lower-income displaced residents could not transport their strong ties after relocation, while wealthier transplants were able to utilize their weak ties, even at a distance.[10] Both A. H. Fritz and C. E. Barton write about the development of "therapeutic communities" after disasters, and the experience of the Bayou Oaks families bears out many of these conclusions.[11] For example, preexisting ties, even weak ties, between flooded residents and local organizations paid off when those ties were activated.[12] Those ties were powerful in terms of providing assistance, because the helpers had qualities in common with the recipients, especially sharing a social class.[13] In addition, the resources that their social networks offered "collectively . . . represented a source of group power that low-income communities lack."[14] The Bayou Oaks families enjoyed a class-related "resource depth," before and after the flood, which they were able to draw upon when they needed assistance.[15]

Some friends and family members, however, offered help to the mothers that was not actually helpful, or that required the mothers to expend more labor. For instance, some grandparents offered to keep the kids for a few days or to acquire supplies like a dehumidifier or boxes, but then expected the mothers to stop what they were doing at the house and drive halfway to meet up for the exchange. As one mother put it, "We don't even *have a car* right now!" Other relatives, trying to be thoughtful, shipped new toys, clothes, and books for the children to homes that had nowhere dry to store anything, and where the families were trying to get everything *out* of the house. One out-of-town sister, while on an unfortunately timed tropical vacation with her boyfriend, offered to send her estranged ex-husband, who lived nearby, over to help, an idea that appalled her flooded sister. And in one memorable instance, an elderly aunt shipped cases and cases of tampons to Rebecca, who sat in her ruined house and opened the gift-wrapped boxes from Amazon, puzzled and now surrounded by thousands of Super Plus tampons. Each gift receipt said only: "Rebecca: Enjoy your gift!" By the time of our interview a couple of weeks later, Rebecca was able to laugh about it, sending out a group text to her neighborhood friends, who had by now heard the story about the tampon incident:

> So, like, do y'all think I should donate them to a shelter? Or should I just walk around the neighborhood handing them out? Like, "Hi, you don't know me, but I see you have a pile of sheetrock in your front yard. May I offer you some tampons to help you through this difficult time?"

Another phenomenon that irked the mothers was when people who'd never been flooded personally offered platitudes about "what was really important," such as the health and safety of the family. The mothers understood their friends and family were trying to be helpful, but nonetheless perceived the statements as minimizing what the mothers were going through. One mother who'd been through this twice summed up what many mothers were feeling:

> And what's frustrating in these situations—and it's . . . there's a group of people that I call well-intentioned assholes—because they mean well . . . but it's not helpful. And people go, "Well, it's just stuff." And that's fine too. I do not care about the stuff. But I still have to deal with the stuff. I still have to catalog it. I still have to pick it up to throw it away. I have to dry it. I have to store it. I have to box it. I have to smell it. I have to rip it up. I have to haul it to the street. I have to discuss it. I have to write it down and I have to look up a value for it. I have to deal with it. There is so much physical labor involved in these things.

While on the whole, the mothers were very grateful for the help they received, it was not an unalloyed good. The helpers sometimes said hurtful things or offered assistance that just made the mothers' lives more complicated, but which they might feel obligated to accept nonetheless. The grief they experienced from the flood was compounded by these complex emotions during this early recovery period. In addition, the mothers were also confronting a sudden, violent loss of privacy, which they were not accustomed to and did not welcome.

In order to accept help in the immediate aftermath of the storm, mothers had to throw open their front doors and invite people into their private lives. Not their performative lives, but their real lives. Not just the entertaining spaces, the living rooms, the kitchens, and the back yards, but also the private spaces: the bedrooms and the master bath, now ransacked by flood waters. Many of the mothers struggled with the way the flood made what was normally backstage, their interior lives, front stage.[16] These families were used to a great deal of privacy and control over which aspects of their lives and homes they would share and how, and the moms usually saw to it that the semipublic spaces in their homes, the rooms that friends might visit, were at least tidied up before people came over. This was part of their identities as good mothers and wives and part of their curation of their home lives.[17] The flood changed all that. As Tracey put it:

> Now everyone can see all of our house with all the—you know—insides turned inside out. It's kind of like when you give birth and everybody comes through your hospital room. And you look a mess, and you don't know who's looked at what. It's kinda like that. So you know 'cause we were one of those like very private people who didn't really wanna open our house to a lot of people. And then it was just like flung wide open.

While the mothers counted many friends and acquaintances among their neighbors and the other school families, they were not necessarily open books with each other. Even in their moment of true need, the mothers didn't want to "look a mess." And while the mothers were grateful to Bayou Oaks Elementary for sending teams of teachers and staff to their homes to help out, it also created some moments of awkwardness. One mother actually covered her face with her hands as she described the school work team descending on her house:

> The mucking team was a team from Bayou Oaks Elementary. The principal, and the nurse, and the PE coach . . . they were all there. It was crazy. I was like, "Oh, the principal's here." I was like, "This is wild." I mean, you feel bad, because, like, your house looks terrible. And, like, hmm [embarrassed noise]. First time you come into my house, and it looks like this . . . and you just have to go with it.

It was clear that the mothers, as guardians of upper-middle-class stability and the home, were worried about their housekeeping, or their clutter, or their décor being judged—even by kind and well-meaning volunteers. Even weeks later, in our interviews, many of them still acted mortified at the memory as they described the scenes, cringing, closing their eyes, groaning, and covering their faces as they recounted the experiences.

Some of them had had the presence of mind to try to preempt embarrassment as the volunteers descended. Amy, who was a first-time flooder but had helped clean out other houses after the first two floods, was struck with an epiphany right before helpers started arriving:

> The other thing I was joking with Mark about too was, it was like, "All right we've got about an hour before everyone is gonna show up here. If there's anything you don't want anyone to see, now is the time to take it. Put it somewhere, because they're gonna open all the shit. Like they're gonna open all the doors, all the drawers. We are not gonna be able to stop them. So I know what I'm going to grab. You need to think about this."

Amy reported that in this instance, Mark was, as usual, appreciative of her forward planning.

To walk the sidewalks in Bayou Oaks during this period was to walk through a gorge, with drywall, furniture, carpet, sodden books, and teddy bears rising around you instead of rock. The debris rose twenty or more feet in the air. While trucks from the city quickly began to pick up the piles, there was so much damage across the city that it would be months before the curbs in Bayou Oaks were clear of debris. During the communal cleanout process, across the neighborhood, "scavengers" were picking through the piles of debris to find things to take or sell. The mothers did not appreciate this, for two reasons. First, these were people from outside the neighborhood, invading their domain and taking their things. These were not the sort of people who were usually in the neighborhood, and it felt shocking to the mothers to see them there. It was as though their things were being sucked into a completely different world, a world the mothers saw as scary or dangerous. Even though they viewed their belongings now as trash, the idea that someone might take their bookcase or chair still felt violating. Second, they could not bear the thought that anyone might rehab something that had been contaminated by floodwater and sell it to unsuspecting people. One father went so far as to hack up the family's antique furniture with an axe so that no one would take it and try to resell it later. The families felt a sense of precariousness—economic and social—when their belongings were on the curb. In normal times, that precarity was hidden. The flood forced it out into the open. It took months for all the debris to leave the curbs and for that sense of precarity to begin to fade.

If the debris and remediating phase of the disaster looked similar across Bayou Oaks, the next phase of recovery—financial—did not. Although the neighborhood as a whole was affluent, there were relatively large differences in both income and assets across the families in the study. These differences would become more apparent, to themselves and to others, as families made decisions about what to do next. And just as the mothers had to give in—with gratitude but also with embarrassment—about accepting the volunteer labor available to help clean out their homes, they also had to make complex and emotionally fraught decisions about what sorts of financial help to seek and to receive.

The storm's aftermath presented mothers with short-term and long-term financial stresses. In the first phase after the storm, the families had to find and pay for temporary accommodations, secure a rental property, pay for fans and dehumidifiers, purchase new cars to replace their ruined ones, and in some cases, pay laborers to help with the cleanup of their homes. This was all on top of having to continue to pay the mortgage for a home they could not live in. The financial strain of the storm and its destruction hit the mothers right away, especially those who did not have much in the way of liquid assets. These tended to be mothers in the bottom half of the household income distribution in the study (those with household incomes less than $180,000). They might have significant assets saved for retirement, but that didn't mean they had large savings accounts or other ways to access cash. One mother echoed many others when she said that her family did not have a big financial cushion they could easily access for their day-to-day expenses:

> So I'm not saying we're destitute. But we have no cash flow. We may have a little bit more in the bank in IRAs and in retirement funds. But

we don't ... we're not destitute, but we are in no way, shape, or form well off. When you look at the whole picture, it might look better than it is on a daily basis. You know what I mean?

Many mothers, like Amy, described tightening the purse strings in the immediate aftermath of the storm: "And so you just put [all spending] in lock-down, because you gotta have cash on hand." Affording rent on top of the mortgage payment was feasible for some of the mothers but not all. One mother who lived in a furnished home donated by a friend during her remodel didn't understand how people were managing:

> We were just financially not in a place. . . . And I don't know who is. I don't care how much money. . . . I don't know how people can afford a mortgage, and rent, and then this crazy gray area of "Am I rebuilding? Am I raising? Am I gonna have to take out loans? Am I gonna actually get the insurance money that I feel like I deserve?" And . . . you know that extra layer of . . . of stress. I mean I don't know how people afford it.

None of the Bayou Oaks mothers were seriously worried about having their homes foreclosed on in the storm's aftermath. In most cases, they had good, steady jobs with benefits (or their husbands did), and their household incomes were well above average. But like many middle- and upper-middle-class Americans, they did not have significant savings on hand; they still struggled day to day after the flood and were put in the difficult position of having to accept help from friends, family, and even strangers. This was not a comfortable feeling. Even though they had been through a terrible trauma, some of them multiple times in just three years, the mothers were eager to distance themselves from people who were truly at risk and truly in need of charity.[18] This attitude came partially from a place of wanting symbolic social distance from economic precarity, but also because the mothers were firmly aware of their economic privilege relative to many others in Houston and experienced feelings of guilt about accepting help because of this awareness. These feelings of unease about their status relative to the rest of the city added an additional layer of discomfort to their dislike of having to accept help.

Although it was apparent to the mothers that they needed to accept help after the storm, both in-kind and financial, most of them still struggled with the concept. They were not used to others helping them, except with in-kind assistance like carpools and playdates. Their identities as self-sufficient, capable upper-middle-class families were threatened by needing assistance. All of the mothers ended up accepting help of some kind, although the amount of help differed considerably from family to family. Several of the mothers expressed discomfort with accepting help because of their own identities as capable, smart women. Rebecca gave a typical response:

> I feel a lot of safety in the fact that I'm smart. And I think that in any situation I can think, and think, and think, and think . . . and everything will be perfect. And with this kinda stuff, that's not true. And that's because I feel . . . I've never not been capable. I've never not been able. So, um, the idea that I would have to face that I'm not capable or not able is difficult for me. . . . So I think I've definitely grown a lot in that area with this flood because I've accepted help [this time].

The mothers took pride in their roles as household logistics managers and in their ability to figure things out. They spent a lot of time optimizing their family's lifestyle and experiences, so it was challenging when even their best efforts could not solve a problem like storm recovery. Michelle expressed feelings similar to Rebecca's:

> I'm not one to ask for help. I don't . . . it's real uncomfortable for me. I don't know why, that's like a whole other psychiatry lesson. I'm sure it's something that stems from I have to be, like, an independent woman. I just . . . and it makes me super uncomfortable to ask anyone for anything. I'm working on it.

Other mothers specifically expressed discomfort with accepting help due to their own financial situations, which they knew were much better than others' in the city. As one mother put it:

> I feel bad taking things, you know. I mean we don't qualify for a lot of financial assistance 'cause we make more than . . . I mean, I drove a $90,000 car. So it's hard for me to, you know, accept people's financial aid because we'll be fine. We're tightening up our belts, and we've always lived well under our means. So . . . it's hard to accept . . . you know. It's just hard to accept. [There is] a sense of guilt, because I know there's somebody else who could use it more. You know? I feel like I'm usually the one that helps, you know.

Even though this mother knew her family would eventually "be fine," they still had tens of thousands of dollars to pay out up front and felt they needed to accept help, something that was unusual for her. There was comfort in the idea that their families helped other families in need, and even if the flipping of roles was temporary, it was not welcomed by many mothers. Samantha put it this way: "I don't know, I just like to do everything myself. I don't know. Just, I like to do more helping others. I don't like to be the one receiving the help."

Helping others was part of the mothers' identities both as good community citizens and as good mothers, so it was challenging to feel that identity being threatened. Although it was difficult, the mothers ultimately rationalized accepting help because they were truly overwhelmed, both by the physical and emotional labor of storm recovery and by the financial strain they felt. But not all assistance was deemed the same by the mothers, and they drew symbolic distinctions about the kind of help that was and was not socially acceptable. First, gift cards allowed friends and family to give (and the mothers to accept) modest amounts of money that did not have the same symbolism as handing someone $50 in cash. Cash seemed like a handout; a gift card was like sponsoring a dinner out or a Target run for the family. Mothers preferred to accept modest financial gifts rather than large ones, and for larger amounts, loans instead of outright gifts; they also rationalized accepting gifts based on their perceptions of the means of the givers relative to their own. Wealthier givers' gifts were more readily accepted. And finally, fundraising online through sites like GoFundMe was a major point of contention in the neighborhood. This took the private gifts mothers were frequently accepting and turned them into public gifts, something visible to others and condemned by some of the mothers.

The mothers had broadly similar ideas about the kind of help that was acceptable and the kind of help that was too much, although each drew her own threshold of acceptability a little differently. That gift cards were acceptable while giving cash was not seems to be a widely understood social norm of upper-middle-class giving, because most of the mothers were completely inundated with gift cards, like Julia's family:

> You know random place is, you know, handing you a $50 gift card . . . so there's, you know, gift cards coming home from [a private school], and, you know, gift cards coming home from Bayou Oaks Elementary. And you know, there's random gift cards right and left. You know, random old friends from college sent me a $100 Visa Card. I'm like what? What?

Because the gift cards were not going to move the needle on their renovations or fund a house elevation, they carried less weight and meaning than a larger financial gift would. The mothers did not have to angst over whether to accept them or how to spend them. They could just appreciate them for what they were: a helpful gesture.

Some of the mothers, however, expressed unease about accepting even these modest gifts because "other people need it more." Several mothers actively redirected the gift cards they received, including Tara:

> But there are so many people that didn't have it, and I just felt guilty for taking this help. And I was like, "We're actually doing pretty good." You know? So that was probably part of it was just the . . . the guilt for me in receiving help when I knew there were other people that needed it more. . . . And so I just passed it along. Because that was our . . . that was our blessing. So every time oh, we got . . . my kids' schools sent home whatever gift cards . . . pass them along.

Even when mothers tried hard to redirect the small gifts they received, they sometimes had limited success. An old friend of Lucy's mother stopped by one day and handed her $100. She insisted that

Lucy keep it, when Lucy tried to resist. Later that day, Lucy passed it along to her friend, who was watching and feeding flooded families' children while their parents cleaned out their homes, to help with her costs. A couple of hours later, her doorbell rang and it was a different friend dropping off a $100 Target gift card: "And I was like stunned, because I was like, 'What?! I just gave that $100 bucks and like within an hour it came right back.'"

The next day, Lucy went to pick up some items from the dry cleaner and got to talking with the clerk, whose apartment had flooded. Lucy ran out to her car and got the gift card, then gave it to the clerk. She drove home and checked the mail. There was a Target envelope, sent by her sister: "And then I go home that night, the mail—checking the mail—and sure enough a thing from Target. I'm like, if this is for $100 I'm gonna scream. I open it up and it's $100! Isn't that crazy? It was like totally cycling."

One thing this group of mothers who were uncomfortable about accepting the gift cards had in common was that they had some of the lower household incomes in the group. It is possible that smaller gifts seemed larger to mothers whose families had less income. It's also possible that these families had closer ties to middle- and working-class people, and therefore more of a sense of what a $50 gift card could mean to a poorer family. Or perhaps their own class status felt more precarious, so accepting a gift card felt more stigmatizing for their own identity.[19] On the whole, however, the mothers were grateful to accept gift cards from friends and family. These smaller gifts helped to subsidize the extra expenses of living in temporary places, which was useful because the families were also trying to find large chunks of cash to get started on their home recoveries.

The Bayou Oaks families were affluent American families by any standard. But many had stretched their home budgets just to afford living in the neighborhood they desired so much. As a group, they were not sitting on a lot of liquidity. While the mothers did expect to relatively quickly receive some rental assistance money from FEMA (typically about $2,000), and most received a standard $400 check from the Red Cross, their flood insurance payouts could be tied up for months or even longer. The payouts were usually released in installments as work progressed, but this was a slow, variable, and uncertain process. One year after Harvey, many of the families were still waiting for the last of their insurance payouts, and in two cases had still not received any money at all. And the upfront costs could not wait for insurance payouts, so they had to come from somewhere. Due to the immense scale of Harvey's destruction in Houston, contractors were in high demand and stretched to the limit, and they could afford to be choosy about the projects they took on. This meant any restoration or rebuilding required an immediate outlay of tens of thousands of dollars to retain the contractor and get the crew started. Even if the families had three or more months of living expenses stashed in a savings account, as some did, that usually wasn't enough to pay even the first installment on this kind of major renovation.

The mothers were able to tap into significant financial resources through their social networks, both family and friends. One mother, who had been flooded once previously, pieced together the cash reserves she needed by calling a wealthy client she was close to:

> One of the things that really slowed us down on the last flood was cash, because yes everything was ultimately covered [by flood insurance]. You know the construction, the content stuff was covered. But you need $20,000 to get started. And so this time I was talking to one of my clients, 'cause my clients in general are fairly wealthy, and there's one that I'm really good friends with. And I called him—or he called me—and he said "I'm really worried about you, what can I do?" I said, "I may need to borrow some [money] for two or three days at a time." I said, "At the beginning I might need it for a couple days. Later on it might be for a month. Are you okay with that?" And he said "Absolutely." So there was that kind of relief.

This mother used these short-term loans to fund her contractor; she also used them to fool the management at the apartment complex where she was trying to rent an apartment. With all of the displaced families in Houston, the complex could afford to have high standards to approve new renters. They insisted that she show proof of $10,000 in cash reserves to rent the apartment. So she borrowed the money for twenty-four hours, showed the bank statement to the complex, signed the lease, and immediately wrote a check back to her client.

Another way mothers tapped into their social networks for resources was through fundraising on social media. About one-quarter of the mothers reported receiving money from a fundraising site set up by a friend or relative (in no case by the mother herself), usually totaling between $1,500 and $10,000. This cash was seen as a godsend by the mothers. While many used it to help get contractors started on their renovations, some of them also used it for a down payment on a new car, to pay rent and the security deposit on their temporary homes, or for furniture and other essentials once their houses were rebuilt or renovated. One mother was grateful for the support generated by a GoFundMe account set up by her sister, but also worried about where it was all coming from, and was concerned about "valid" uses of the money:

> I mean if I added . . . it was a lot of money. I mean it was probably close to $8,000 or $10,000. And it's there but you know we had to replace [my husband's] car and all of a sudden it's like we need a down payment. And that's the money, you know? That's a valid . . . that's a valid way to help us use it that we wouldn't be able to put you know $4,000 or $5,000 down on a car right now, you know.

Clearly, guilt was driving the uses this mother viewed as "valid" for this money and those she would view as "invalid."

Many of the mothers reported accepting these crowdsourced funds only reluctantly. They clearly perceived an undercurrent of disapproval about the practice in the neighborhood, and they felt some awkwardness about the public nature of GoFundMe, which was usually shared on Facebook by their friends. Some of the mothers reported redirecting some of the funds to flood relief groups or to

Bayou Oaks Elementary, or said they had plans to do so. One mother explained her uncomfortable feelings about personal fundraising this way: "Because we don't need like other people need. Like we could find a way. We could make it work. Like it just felt awkward."

Although most of the families needed financial assistance in the first days after the storm, most were acutely aware of their financial standing relative to others and felt they "could find a way" without accepting personal fundraising. While the mothers' discomfort with accepting this cash was usually not something that was openly discussed, one mother had faced direct shaming from a neighborhood friend for accepting the money, and she wept as she described how that made her feel:

> My cousin in California—she just felt really . . . she wanted to do something. She started a fund . . . and she put it online and it just exploded. I mean they . . . it ended up being thousands of dollars. It's been amazing. But then I have my friend who was like, "It's just popping up on everybody's pages all over Bayou Oaks and it's just such a shame. You know that other people have flooded more than once and like it's just such a shame." And I felt very shameful that I had taken money. And to this day I have a really . . . really . . . [crying] I don't talk to anybody about it in Bayou Oaks 'cause I don't know what their reaction will be. I don't know if they'll be mad or. . . . I feel very ashamed about it. I don't know. It felt wrong.

This mother was made to feel as though she was the only Bayou Oaks mother who'd accepted money through fundraising, when in fact a quarter of the mothers I spoke with had. Even though most of the mothers expressed some reluctance or embarrassment about accepting the money, it was true that in the heady days following Harvey, with images of water-besieged Houston all over the national news, people wanted to help their friends in Houston. And this was one straightforward way for them to do so. There is no doubt the money was helpful as the mothers struggled to put together the cash they needed to replace their belongings, buy a new car, and find a temporary home.

While some of the mothers who received GoFundMe money did not report any discomfort with the idea, most of the mothers who did not accept fundraised money were actively judgmental about the practice. The disapproving mothers had a couple of narratives about why they did not think it was appropriate to use crowdfunding. First, they believed that insurance should be enough to pay for recovery, and that it was inappropriate for those who had insurance to ask for extra money. And finally, they believed that those donations could be put to better use elsewhere. Even though the process of flood insurance payouts was not smooth for most of the families, many of the mothers nonetheless believed that the insurance should take care of the major financial needs. It was gauche, then, to ask for more money via fundraising. Many mothers said something like the following:

> Like it actually makes me uncomfortable having GoFundMe accounts for them. I feel like someone sent out that for, like, [another local family] and I was like, "That's not right." Because they have insurance and they. . . . I mean, like, I know like we've all lost stuff or whatever. But like it feels . . . it feels weird to me to ask for people for money, when you have insurance. You really should be all right with your insurance. And so that bothers me.

Michelle reacted strongly when asked whether she considered having a GoFundMe:

> That would be like my worst nightmare. I . . . I mean. . . . That means I'm like, desperate, and I don't know that. It's for people who really need it. I feel like, you know, we're fine. Like I might be in debt for the rest of my life, but we're okay. It's hard for me to, like, receive the gift cards and stuff to be honest. I feel like I want to donate them back to somebody who like really needs them, you know.

For Michelle, appearing needy or desperate was her *worst nightmare*. It was something to be avoided at all costs. Her house was ruined, and all her waterlogged things were out on the curb, but she wasn't interested in asking friends for financial help.

Another arena where mothers felt discomfort was large financial gifts or loans from family. In contrast to the quasi-public nature of the GoFundMe fundraising, gifts from family were largely invisible to the rest of the neighborhood. But this did not mean they came without baggage. In total, ten mothers received large financial outlays from family members, most in the form of checks, but also via alternative avenues. The financial schemes they relayed were complicated, strategies only people of means would know about and know how to execute. In addition to helping the flooded families, the strategies had tax advantages for both the giver and the receiver, as described by one mother:

> But then [my husband's mom] was talking to him, and she was like, "What we want to do is [purchase] half your mortgage and then pause your mortgage payments indefinitely." So it's, like, a way to get around, like, taxes, because with income there's a gift rule, like you can only give like, $20,000 a year to people or whatever. So this is a way around it where they . . . they basically become our mortgage company, but don't make us make payments. So it's like basically giving us like an inheritance now, when they're alive.

Another mother had two different relatives offer to purchase the family a new house outright but found the implications of accepting such a large gift to be complicated:

> And so unfortunately, where I am like overwhelmed with gratitude, and emotion, and thankfulness—and saying, "Thank you . . . where are we moving and when"—[my husband]'s like, "I don't think so." Because, you know, "That makes me feel like a failure." So there are a lot of intricacies.

Another mother who rejected a large gift from a family member did so by employing a complex calculus about what could be accepted and from whom:

> We had a relative send us two checks for $36,000 each. And we said, "Thank you very much but no thank you." It was very nice, and she has

the means to do it. I think that there's, like, a ratio of how close you are to someone and how much money they send you. And you cannot exceed ... do you know what I mean? Like if your mother gives you half a million dollars, and she has it to give, then you say, "Thank you so much, I really appreciate it." If like someone who is far away, who's a second cousin by marriage once removed, and they send you $10,000, that can feel really different.

While the mothers who received large loans or financial gifts from family members were grateful, they also recognized that intermingling finances with family members was a delicate proposition. In addition, they recognized how fortunate they were to be able to rely on family for substantial help.

For most of the families, the largest financial outlays they were to receive came not from private sources but from public ones—specifically, their flood insurance companies, underwritten by the federal government. Unlike their feelings about private sources, the mothers experienced no qualms about accepting this help, even those who had received large flood insurance payouts once or twice before. From their point of view, they were entitled to these payouts because they had paid (often high) flood insurance premiums. The mothers therefore applied their organizational skills, their facility with bureaucratic processes, and the tenacity they ordinarily directed toward their family lives to working with FEMA, flood insurance companies, and mortgage lenders. As a result, many mothers were able to get more in flood insurance payouts than their families were originally offered. They accomplished this through extensively documenting their losses, their ability to stay tightly on top of their adjusters, their willingness to "throw down" when necessary, and their tenacious but purposefully charming behavior toward the adjusters. Although most mothers reported a high degree of frustration working with their insurance and mortgage companies, they were nonetheless able to apply the skills learned from navigating other bureaucracies to ensure what they considered fair payouts for their families.

The ability of families with more resources to secure higher insurance payouts has been documented in other postdisaster studies, including Rebecca Solnit's *A Paradise Built in Hell*. Noting that these families do not act with the intention to take money from other disaster survivors, Solnit wrote, "Even when malice is absent, middle-class people who maintain extensive documentation and are good at maneuvering through bureaucracies do better at getting compensation."[20] After hurricanes Andrew and Katrina, researchers noted that "system skills" were both useful in obtaining postdisaster assistance and concentrated among those with higher incomes and educations.[21] In Bayou Oaks, while both mothers and fathers worked together to document their losses and maximize payments, the mothers were typically the ones who interfaced with the adjusters. They were used to being the person in the family who applied "system skills" to maneuver their families into desired school, religious, and extracurricular activities and to navigating the workplace and institutions of higher education.[22] They simply successfully transferred those skills to a different scenario.

Flood insurance on the structure of the home was required by the families' mortgage companies in the Bayou Oaks area and had a maximum payout of $250,000. Contents insurance, or insurance on household furniture, electronics, and other items, was optional, and the maximum payout for contents insurance was $100,000. Only one mother who had flood insurance did not also have contents insurance, although some of the repeat flooders had learned to add contents insurance the hard way, after their first flood. In the first phase after the storm, the mothers and their husbands worked diligently to document their losses, because they knew it would result in higher insurance payouts later on. The documentation was useful for structural payouts too, but mostly, the families worked to max out their contents payouts, which they had more control over. This maximization was important for several reasons. First, the mothers felt entitled to the maximum possible payout because they had paid for the insurance every year (on average, flood insurance in Bayou

Oaks prior to Harvey cost $4,000 per year). Second, to refurnish an entire house was expensive. And finally, many of the mothers used the cash from their contents payouts to cover other costs during the recovery process, like rent for their temporary homes.

While the Bayou Oaks families were eligible for some minor financial assistance from FEMA, most of the money they would end up receiving came through the flood insurance process. The mothers detailed a plethora of woes related to dealing with their flood insurance claims. One mother, a full year after Harvey, was still waiting on her final $30,000 payment from insurance, just because the bank holding her mortgage had been bought by another bank. Another couldn't get her adjuster to call her back, because he mistakenly thought her family was suing the insurance company. Checks were held up due to one missing signature on documents dozens of pages long. Mothers faxed paperwork multiple times, only to be told the companies had not received it. In all, most of the mothers spent hundreds of hours on paperwork, documentation, and phone calls with their adjusters and banks. In three cases, the mothers' husbands were the ones who did most of the haggling, but for the vast majority of families, just as for most of the logistical and organizational labor prior to the flood, it fell to the mothers to do this work. It was clear that the year following the storm had been one of intense frustration and stress on this front. A year later, some families were still waiting for checks from their insurance companies and banks. Many mothers reported having to raid savings accounts and retirement funds to pay the installments their contractors demanded in the meantime.

Mothers who had been flooded and had been through this process before tended to fare better, partly because they had recent receipts for their earlier renovations, and partly because they knew what to expect and had flood capital they could deploy. Despite knowing what to do, and generally having an easier time getting their payouts than the first timers, having to deal with FEMA and the insurance adjusters again was not something any of them relished. Nancy, a three-time flooder, was not amused by FEMA's attempts at good customer service: "And here's the worst FEMA story of all. When I reapplied that night—you know as we're getting ready to flood I reapplied online—I get a text message: 'Welcome back, Nancy.'"

Meghan, a twice-flooded mother of two, spent a considerable portion of her time at work haggling with her insurance company and bank. She even pressed her assistant at work into service to help with the back and forth. Even with all that effort, it took months before she received any insurance money: "I, seriously, was, was going come out of my skin. I was like, 'What is it going to take before you people give me my money?'"

Laura, who did not work outside the home, also spent an inordinate amount of time on the bureaucratic process and felt that the system was probably constructed to be difficult on purpose:

> But knowing what we had to go through to get to that point, it's—it very much feels like a purposeful beatdown. Like, how tenacious are you going to be? How bad do you want it? "We'll give it to you if you really want it, but if we can get away with not, then we're good with that too."

Sally, a three-time flooder, was still getting payouts for her second flood after Harvey: "And we were still getting—we actually got a check from our last one after we flooded in Harvey because we were still dealing with fighting the insurance company."

Even though the Bayou Oaks mothers had bureaucratic skills and spent hundreds of hours preparing their claims, they still experienced an enormous amount of frustration and red tape before they got their insurance payouts. But they were nothing if not tenacious. Melissa, like many of the mothers, described how she carefully vetted what the insurance adjuster was proposing for their payouts and figured out how to substantially increase it:

> So what I did is when I got the proof of loss, I went through it like with a, you know, finely toothed comb, and I had our proof of loss from the last flood, and I, like, made sure that they were giving us the same amount for different things. . . . And so, I went through with that and

I went back to the adjuster and I was like, "You know, our—all the doors are custom, you know, all the cabinetry was custom, you know, you need to make those changes." And so, I made him go back through and change everything. And then, he had missed like a couple appliances, and so I would like, just list out everything that he missed—and so we ended up getting $30,000 more back because of that.

The final strategy the mothers employed to maximize their insurance payouts was to be solicitous of their adjusters. Because they knew that the scope of flooding meant their adjusters were extremely busy, they also recognized that being nice would get them further than being difficult. And they hoped that being easy to work with would even move them ahead of the other clients the adjuster was working with. The mothers also recognized that this was a valuable skill they had cultivated over the years and could deploy as needed. As Sarah said:

Like, I do know how to talk to people, and the insurance adjuster came, and I just—I didn't, I didn't want to lay it on too thick, I didn't want to do anything [untoward], but I just—and this is how I've always been, this is why I'm in P.R., it was just—I developed a relationship with him. I made it a point to understand that this dude didn't sleep for literally six weeks. So, I needed to be the person that was like understanding—and not crazy.

The mothers clearly understood that although at times they wanted to call their adjusters and scream in frustration, they needed to be on the adjuster's good side, as one put it: "And—but it's one of those—you can't hound the adjuster because there's one—you can't be rude to them, and, you know, complain because they're the ones giving you the money." The mothers recognized the power dynamics in the situation and made sure they were well-behaved to their adjusters. They used skills learned in higher education and on the job to organize, document, and submit their losses to their insurance companies. In all, the mothers' organizational and social skills were critical for ensuring the maximum payout from their insurance companies.

The Bayou Oaks families received an enormous amount of assistance after Harvey, including in-kind labor and donations, small and large financial gifts and loans, and flood insurance payouts. All of this assistance was facilitated by social class. The families had friends and relatives who could take time off work to help with recovery, who had access to financial resources to share, and who could help care for children. And the mothers had bureaucratic, logistical, and social skills they could use to ensure maximum insurance payouts for their families. This outpouring of help in Bayou Oaks helps to illustrate why disasters may actually increase inequality in communities. The Bayou Oaks families, as they were well aware, were already well off before Harvey. While they experienced a terrible tragedy when their homes were ruined, they were besieged with assistance from many fronts. All this help, and all of their skills, however, were not enough to prevent widespread emotional and financial difficulties, which would continue to plague the families for more than a year.

5 Family Impacts

THIS PAST YEAR HAS REALLY BEEN SO WRETCHED

At our first post-Harvey interview, Ashley, principal of Bayou Oaks Elementary, had seemed weary but resolute, firmly in control of shepherding her school community through a disaster. A year later, she seemed to have more energy and she had regained a sharpness, somehow—an edge. The year had changed her, clearly—a year of managing dislocated children and teachers, satisficing everyone. Now, in early November 2018, she was roughly two months into the second school year post-Harvey. Ashley was in a reflective mood. She was preparing to lead a Bayou Oaks delegation to Panama City, Florida, where a school had just been destroyed by Hurricane Michael. The convoy would bring supplies and treats for the children and teachers, and most importantly, some firsthand experience in recovery and school community reconstruction.

About eight days post-Harvey, Ashley was ripping out drywall in a family's house with a few teachers when she got a phone call from someone on the school board. He threw out the suggestion that there was an empty school building that might work, located a couple of miles north. He offered to take her to see the place the next day. Ashley agreed and then, not saying a word to anyone else, dropped what she was doing and raced to the campus alone in her SUV. "I was literally there in less than five minutes." She bypassed the "no trespassing" signs and walked the entire campus, peering in the windows and doors. Her first thought was that there was no way this was going to work. It was too small, too run-down, and too disgusting. Wires hung from the ceiling, mold was visible on the walls and carpets, and cockroaches scurried across the floors. Unwanted furniture from across the district was crammed into every room. There weren't nearly enough classrooms. She didn't want to ask her teachers to come to this place, much less seven hundred young children. When she relayed her concerns to the district, they assured her that they would get the site cleaned up. Ashley dubiously imagined some guys arriving with bleach wipes and Swiffers. That wasn't going to cut it. Days passed without any crews showing up, and Ashley became increasingly frustrated. Eventually, to her relief, the district sent in the promised, massive crew of people. They immediately began ripping out carpet, repairing the network, clearing out debris and broken furniture, and doing their best to make it habitable, working 24/7. She started to believe that the new location might work—but she did not want anyone else in the school community to see it in its current condition. It would be too upsetting to envision their children in the dilapidated place, too much for them to handle after what they'd already been through.

Ashley was working eighteen-hour days, seven days a week, rushing back and forth between the old campus ("Bayou Oaks OG") and the new campus ("Bayou Oaks North"), visiting the district headquarters, doing media interviews, meeting with the principals from other flooded schools, visiting her teachers, and communicating with her parent community. Once she knew for certain that her school had a home, even if it currently wasn't fit for use, Ashley sent out an encouraging message to the school community, which downplayed the issues at the new campus and sought to reassure parents. The message also broadcast the new start date for Bayou Oaks North: September 25, almost a month after Harvey and two weeks after the rest of the district returned to school.

As the work progressed on the Bayou Oaks North campus and the start date crept ever closer, Ashley began to be concerned that it would not be finished in time for her teachers to get their classrooms properly set up. Ashley was livid; she called the district and gave them a piece of her mind: "I cannot bring my teachers to see it like this! If this is what they walk into they will lose their minds! They *will not* make it!" The fragility of her teachers, who had lost all their classroom materials, was her primary concern. She was also managing the expectations of parents; she was irritated when some mothers snuck over unannounced to inspect the school. "Don't do it!" Ashley recalled thinking, with clenched teeth. "I didn't want them to be nervous. I wanted them to see it when it was ready." Ashley's patience was fraying. Her stress levels since the storm had been extreme, and although she was businesslike and focused, she sometimes let her guard down just enough to marvel at some of the situations she was handling at all hours of the day: "So I had a few moments, where I felt, you know, 'This is a little outrageous.' You can tell yourself that this is outrageous, but you're gonna have a decision to make in about two minutes, so you gotta pull your show together."

Once they finally got into the building, the teachers were immediately confronted with the limitations of the campus that Ashley had been living with for weeks. They did not hesitate to voice their concerns. Ashley admonished them to remember to be grateful that they were all together instead of split up across the district, to stay positive, and to work politely around the crews who were still installing smartboards and cleaning up. Ashley knew how far the campus had come in the last two weeks; the teachers did not. The teachers, needing more hands, had the PTO send out a request for parent volunteers to assist with classroom setup that Friday. Parents and teachers worked side by side unboxing supplies and organizing books in classroom libraries. Brand new classroom rugs went down.

Posters were hung on the walls. The teachers labored on through the weekend. The campus went from zero to a fully functional school in three and a half days.

By Sunday afternoon they were ready to receive families for the meet-the-teacher event. Before the deluge of families, Ashley reminded her staff that the parents, also fragile, would be looking to them for positivity. The teachers rallied. Out on the lawn, the transplanted two-foot-high wooden letters covered in children's drawings stood proudly, reading: BAYOU-AKS. The second O, now presumably floating somewhere in the Gulf of Mexico, would be replaced later that afternoon by a dad handy with power tools. The chain link fence in front of the school was decorated with red solo cups poked through the links to read, "BAYOU OAKS STRONG!" The staff was ready, and families poured into the building, breathing life and warmth into the space. A spirit of celebration and excitement pervaded the air, and to Ashley's relief, even if the parents noticed the building's inadequacies, no one mentioned them. They were relieved just to have a school home. And that spirit of togetherness would persist throughout most of the post-Harvey year. The second year, though, was a different story.

At first, during my second interview with Ashley a year after Harvey, the focus of our conversation was the children and how they were faring in this second academic year post-storm. Ashley felt there were some signs of anxiety in the children, definitely, and plenty of nerves during heavy downpours, but overall, she felt they were doing really well and had settled fully back into the school routine. The mothers, however, were a different story:

> What has surprised me the most is how in hindsight, the first year wasn't the hardest part of this. It's this second year that is the hardest. I think they're [the mothers] so sensitive this year. Everything is a sensitive issue, and I think it's a combination of our [school] culture, you know and I just think the . . . the adrenaline that we went through that I didn't even know existed in human beings is gone, 100 percent gone.

During the first year, nobody had the bandwidth to complain. They were happy just to stay together, even in a substandard school building miles away from their home campus. Just the fact that school had started, and kept happening day after day, was a miracle. They were preoccupied by their own home recovery processes. Once summer came the next year, most of the families (unless they were building a new house) were back in their homes. A kind of normalcy returned, and with it, the parents' high expectations for what Bayou Oaks Elementary should be. Mothers of children in the oldest grade wanted their children to have the same traditions and experiences that prior fifth graders at Bayou Oaks had had, some of which weren't possible in the new building and new reality. These negotiations with increasingly irate mothers consumed many hours of Ashley's time, while what she felt she needed to focus on was placating her increasingly frustrated teachers and staff.

In the second post-Harvey year, mothers not only were anxious about their children's school experience and maintaining the school culture, but they fretted more about their children's achievement and learning. And that fretting led to more work for the Bayou Oaks teachers:

> And I just couldn't figure it out, but I also have probably twice as many kids already, this year, going through what we call the IAT process than I did all of last year, so that's like—it's called the Intervention Assistance Team. And parents coming in saying, "Well, I, I just know they're dyslexic," and they're like fourth grade, and I'm thinking, "We've got that, we probably would've detected that before now." But the amount of kids that we are testing for dyslexia—or parents coming in saying there's ADHD, is just—I—I've—at a rate that I haven't seen before.

Ashley believed this phenomenon was directly attributable to the flood. She believed that heightened stress and anxiety in the families led to increased sensitivity of the mothers about any potential issues their children might be facing. In addition, although for the most part the children did not dwell on their flood experiences, she believed Harvey contributed to an anxiety epidemic: "You know, there's a ton of anxiety, like kids saying, 'Oh, I'm nervous about this,' or 'I'm anxious about this, this grade,' or what-have-you."

While Ashley was picking up on increased anxiety among her students and managing parents with heightened sensitivity about seemingly every issue, she was also trying to nudge her school community back into normalcy. Homework was reinstituted, to substantial outcry, and special accommodations or arrangements began to recede. The reaction to this irritated Ashley because she knew she was reinstating a level of homework that had been the same for years prior to Harvey:

> The stuff at the beginning of the year was just—I have never seen anything like it from parents, and it was almost a—this, just, force of negativity coming from parent to teachers, specifically in fifth grade where I literally called parents, and I'm like, "There has, there has to be a bridge here, we're, we're not—we can't do this all year." And the teachers were upset, and parents were upset, and I, myself, was annoyed that we were having to handle it because it was taking time that I didn't think needed to be donated to something like that.

Ashley felt that putting the Harvey year in the past would be the best way forward for the students and parents (and certainly the teachers). This adjustment proved more difficult than she had anticipated:

> And even in thinking about it, I was like: well, we did just abandon everything we could last year that we wouldn't have normally done. We said no, of course you, you know, you just read however many hours [for nightly homework]. And now, we're back to a stricter framework, because now I hear myself saying, "No, no, that year's over, that year's behind us, don't act surprised, this is how it's supposed to be," you know. I'm like, "Ooh, I'm being too tough." You give up so much control or—and expectations you know, from what it, what it should be.

Giving up those expectations had been an essential part of navigating the Bayou Oaks Elementary community through the post-Harvey year. But Ashley knew she couldn't allow it to continue, and she faced resistance from the mothers as she tried to steer the ship back into normal currents. Given what was going on in the family lives of the flooded students, this is perhaps not surprising. The mothers' own family lives faced difficulty going back to normalcy as well, and increased marital conflict, significant emotional management of children, and their own mental and physical health problems added to their already heavy burdens of family responsibilities. While Ashley might have been ready for things to go back to "normal," the mothers were still struggling to get back to the orderly, happy family lives they hoped would return as they got back into their restored or rebuilt homes.

The mothers' role as household logistics managers worked pretty much as designed before, during, and immediately after the flood, with them taking on the majority of the organizational, logistical, emotional, and cognitive labor for the family.[1] Once the post-storm adrenaline ebbed and the recovery process needed constant managing on top of their typical daily duties, the mothers' mental loads became extreme, because they were already taxed *before* the storm.[2] This chapter argues that the mothers viewed restoring their households as part of their domain, both because they normally managed household functions and because it was bound up in their identities as good mothers. Adding managing a recovery process to their daily lives, while they were also determined to maintain their high parenting standards, was too much. Something had to give. The chapter first illustrates how mothers tried to maintain their high parenting standards, then details the consequences of that for the mothers during this challenging year, which included increased spousal conflict, increased time managing children's anxieties and emotions, and significant mental and physical health challenges.

The mothers' expectations for how their homes would run and how their children's lives would be constructed did not change after the storm. Just as they undertook to carefully curate their children's experience before, during, and immediately after the flood, the mothers undertook considerable efforts to "maintain routines" even in the incredibly disruptive circumstances of the year following Harvey. They believed that keeping life as normal as possible—while maintaining the high standards of intensive parenting—would help their children cope with the trauma of displacement and disaster. They also were personally eager to return to a calmer, more predictable daily life and to restore their "domestic culture."[3] They took on these tasks because they saw them as an extension of their role as household managers, and because doing so reinforced their identities as good mothers. The gendered division of labor in their households was thus reaffirmed, rather than deconstructed, during the recovery period. Just as women see emotion work as part of their "family work," the Bayou Oaks mothers added "home recovery" to their ledgers.[4] These efforts were time and money intensive and took place during a period when the mothers didn't have a lot of bandwidth. The stresses added up, and the mothers began to feel resentful about the way that home recovery had fallen on their shoulders.

A core example of this labor was how after finding a suitable temporary home, mothers then turned their full attention to making it as welcoming as possible, so that their children would feel at home. Anna, whose children had flown out to California to stay with her mother after the storm and never returned to their flooded home, was determined to get the family's rental apartment fixed up nicely before they came home, about ten days after Harvey. With the help of her husband and her extended family, Anna outfitted the children's new bedroom with everything from new bedding to coordinated bedside lamps. This was important, she felt, because she wanted the kids to be excited about where they had to live, instead of sad about the home they had lost:

> I wanted them to understand that something significant had happened. But I really wanted them to be able to keep thriving. And I wasn't sure how they were gonna take not being in the house. And so I thought if we can make the apartment look as welcoming as

possible, they will want to be there. They won't focus on not being at our house in Bayou Oaks. They'll want to be here. And it worked. So when they walked in [to their bedroom] they were like, "Wow!" And then [my daughter] walked in—I have a video—and she was like, "I love my room . . . I love my room . . . this is so cool . . . so awesome!" And I was like—"We did it." That response was . . . that was what I wanted for them. It's important for kids to feel safe. So that was very important to me.

Anna undertook not just the labor of finding a good apartment and getting it set up properly, but also anticipatory emotional management for her children. She expected that they would feel unsettled and sad in the new home, so she worked to ensure that wonder and excitement, instead, would be their primary emotions when they saw their new home. Jessica, mother of two, was happy with her decision to immediately rent a house, so her boys could play in the backyard and have their friends over instead of being in a small apartment. She felt the rental home provided a place for the family to breathe for a little while and to have some stability after losing their home, their school, and her youngest son's preschool: "So my kids have had everything moved. Their schools, their home, everything. So it's just . . . it's just been about getting some normalcy back for them. We've been trying to do that as quickly as possible. And I think we've done a pretty good job of it."

Jessica, like Anna, prioritized restoring normalcy for her children in a timely fashion and felt she had done a good job. In order to find a rental home in the post-Harvey period, Jessica scoured online sites for hours and visited several houses before selecting one. She had many other things to do during those first days of recovery, but finding a rental home for her children's sake was paramount. The desire to get their homes—and more-predictable family lives—back quickly ran through every interview. A few weeks after Harvey, the mothers were still optimistic that would happen—and that they could *make it* happen using their logistical skills. However, Ruth, a flood veteran, cautioned against rushing to remodel and move back into an existing home. She was sympathetic to the feelings of first-time flooders, but having been through this several times already, she knew it was smarter to take a little time and figure out what made the most sense, long term:

> It's like they're just . . . they're just grasping for their home again. They want to be back to normal. They want it to be [like it was] pre-flood. You know they want all this stuff. And so with that, the home means something to them. It represents something. And so they just want to get it back. But you know this is gonna happen again. I don't feel hopeful that it's not. Like that's not my perspective. Clearly, I have good reason.

The Bayou Oaks mothers were "grasping" for the stable home life they'd so carefully curated and trying to hold their families together.[5] They planned carefully how to get their children settled in a temporary home where they would feel comfortable as quickly as possible. This labor came at considerable cost, both financial and emotional. And that labor did not result in perfect family routines right away.

The mothers also expressed considerable dissatisfaction—with themselves—over the inevitable disruption to their children's routines. While they knew the cause was Harvey's destruction, they nonetheless blamed themselves for not being able to keep up with their own high expectations. This emerged most frequently as a self-rebuke for not being able to maintain their children's prior extracurricular commitments, as Maria suggested during her initial interview, about six weeks after Harvey: "So this semester's kind of an outlier. We'd probably have them doing a little bit more if mommy had any more capacity for it." Here, Maria, a full-time working mother, explicitly blamed herself for her two children not participating in as many sports and dance classes as usual: "mommy's" capacity was just too low after experiencing a life-altering home destruction. Similarly, Ellen expressed regret—and took responsibility—for her son having to step back from his gymnastics training after his gym flooded: "Once all of this happened, they moved the practices really far out [in the suburbs]. And then we [she and her son] just couldn't get it together in any way, shape, or form to get him to [the suburbs] to go work out with the team." Ellen, also a full-time working mother, "just couldn't get it together" during the recovery period to maintain what she felt was expected of her.

The mothers also experienced regret and self-blame for the looser parenting that had occurred during the weeks of chaos that immediately followed the storm. They remembered how that time had been out of their normal parenting structure, and it made them uncomfortable. As Amy described:

> Because so many... the boundaries were just off. Because everybody... that was like the second week after Harvey. There was no school. A lot of people were still ripping stuff out of their house. And I was at the house all day doing that. And so it was sort of like, all bets are off. All of those things, like, maybe you were that kind of tight parent, who was like, "This is our rule and structure." Things like, "You can watch iPad all day long as long as I can get this thing done." So it changed things.

Time after time, the mothers, with obvious discomfort, said that those weeks had been anything goes, with their kids being dropped off for entire days at the homes of friends. They fretted over unmonitored screen time and the lack of a regular nap schedule and worried about what constant access to junk and fast food would do to their children's diets. The changes had been necessary, because of the recovery labor they had to do, and because most of the mothers didn't want their children around their flooded homes, but this bothered them, and those feelings lingered. That the mothers felt the jobs of creating a temporary home their children would like, maintaining their extracurriculars, and maintaining routines fell to them is not surprising given that mothers usually feel responsible for home and family life, and because these specific recovery tasks are strongly related to the emotional labor of motherhood.[6] The mothers felt responsibility for ensuring their children came through this time of disruption as smoothly as possible. In the follow-up interviews, they were relieved to be largely back in their routines, extracurriculars and all. The time of disruption and chaos had left its mark, and they were happy to leave it behind. Not so easy to leave behind, however, were the consequences of this year of heightened stress and the demands of intensive mothering.

During and immediately after Harvey, the couples in the study largely pulled together. During the initial interviews mothers generally reported being on the same page with their spouses; the mothers may have been making the primary decisions about preparations, evacuations, and cleaning out the flooded houses, but their husbands were generally active participants and helpers. As time passed, however, and the disparities in the recovery division of labor became more extreme, the relationship harmony broke down. In the follow-up interviews the mothers reported significant conflict within their marital relationships. It was often the very first topic they brought up, unprompted. For some couples, conflicts were not about the big decisions, but rather about the routine stresses of daily family life, exacerbated by the families' temporary housing situations.[7] Displaced families were often living farther away from workplaces, schools, and the kids' sports fields or dance studios. More time in the car, especially in Houston's notorious traffic, increased tensions. In addition, the families were all living in smaller quarters than they were accustomed to, with less space to spread out and decompress. Many of the mothers, like Tracey, attributed the relationship conflict they experienced over the year following Harvey directly to where they were living:

> Oh, yeah, there was lots of conflict. Yeah, there was lots of conflict. There was a lot of—and just being in the apartment was very stressful with kids in there. It wasn't terribly small, but it was just being in a place where you have neighbors all around you really close to you, and then your kids are like—the kids were loud —and that business, or just being woken up at 3:00 in the morning because somebody's

walking across the floor above you, that kind of thing. It was so hard. But, yeah, with my husband and I, we were definitely—it was definitely a strain for sure.

The Bayou Oaks mothers thought they had left apartment living behind and were not happy to return to it. They were used to having room to spread out and to being able to turn kids out into a safe backyard, and they felt added pressure to keep their children quiet so they wouldn't bother neighbors.

Many displaced mothers complained about the loss of privacy and felt it was a source of strain in their marriages. Amy, for example, recalled the difficulty of trying to have serious conversations or arguments with her husband in a small space where children or other family members could overhear: "It was definitely stressful. And living with my parents—you can't really fight, you whisper fight, you know, over things." The mothers were used to curating all aspects of their home environments, including where and when they could have conversations with their husbands. A sudden lack of privacy added stress every day. Their temporary spaces also meant logistical changes around commutes and school drop-offs that increased marital conflict across the recovery year.

The central relationship conflict described by the mothers in their follow-up interviews was over the division of labor during the recovery process. In good times, before the flood, the families were mostly free from major financial concerns. Their kids were enrolled in a good school. Their neighbors were friendly. The mothers were able to focus on the day-to-day, typical challenges of balancing work and family life and to run their homes according to their values and preferences. In that scenario, they took charge of the family logistics with gusto, and while they occasionally would roll their eyes at each other about how their husbands had no idea how much they juggled, they were largely content, or at least resigned to the situation. Like many other couples, even those with egalitarian gender ideals, the mothers generally took on the myriad of responsibilities for family life but did not necessarily view that unequal distribution of household labor as unfair.[8] Every mother reported that during "normal" times she was the logistics manager for the household—in other words, the person who did the planning, did the research, made decisions, and was responsible for family life.[9] And in the year following the flood the mothers still had all those responsibilities *plus* those of managing the remediation and rebuilding of their homes, usually carrying out this work from a temporary residence. While sometimes after disasters gender-based power struggles emerge that may ultimately gain women social value and improve their sense of self, in Bayou Oaks the mothers felt intense resentment, anger, and stress, which caused a multitude of problems over the recovery year.[10] Women's doing a lot of unsung labor after a disaster is not new.[11] But in Bayou Oaks, the divisions of family labor that made sense before the storm, and with which the mothers had been mostly content, broke down under the strain of disaster recovery. Because the home was considered to be within their domain, and the home was what needed repair, it "naturally" fell under the mothers' responsibilities. Their experiences show what happens when women can no longer accept the gendered division of labor within their homes and start to push back when it becomes unbalanced beyond recognition, such as after a disaster.

Many of the mothers reported some version of the following: "I was managing everything. I was having to do all the logistical stuff of where the kids are gonna be, and where are we staying, and getting the apartment set up and all that." This mother, like virtually all the others, was responsible not just for managing her home's recovery, but also for arranging childcare and finding temporary housing. Another put it this way, after telling me she had not experienced a lot of conflict with her husband: "We've been pretty agreed about what's gonna happen [with the house]. But right now mostly the work of it falls on me. And I'm tired." An agreement on what to do with the flooded home was one thing, but someone has to execute that agreement. Someone has to meet contractors for bids, find financing, wrangle with the insurance company, check the mail at the flooded home, and monitor progress. And in the vast majority of cases, that someone was the mother—and this did not differ based on whether she worked full time, as two-thirds of the mothers did.

Mothers' reactions to being tasked with the recovery logistics on top of their normal work and family responsibilities varied. A few accepted the arrangement as the most efficient division of labor, because their husbands were used to them managing the home life. Julia, for instance, who worked part time before Harvey but ultimately quit her job during the recovery, explained her management of the recovery process as part of the overall way her marriage functioned, which she saw as integral to the kind of family life she wanted:

> So, I think I need to navigate the rest of it to make that world easy for him. I always tell him, "You don't understand how easy I make your life." But I feel like it's my job to make his life easier. That sounds so fifties housewife. Oh, my God, it sounds so fifties housewife, but I think that there's—something to feeling that it's—it's a lot of pressure for me to navigate the whole—just keep the tone of the house, you know, where I want it to be, where I need it to be so that everybody is happy, and healthy, and going to school, and doing their thing, and doing what they need to do.

Julia saw this work as part of her role, so that she could make her husband's life "easier." She also saw herself as responsible for setting the entire tone of the home. She recognized this as being a rather traditional arrangement, especially after she quit her job, but also felt that her efforts had gotten her family back into their home quickly.

Deborah described a similar division of labor, in which the daily minutiae of tasks associated with building a new house fell to her as part of her role as a mother:

> What I have to do now, unfortunately every day, is like manage the general contractors doing stuff. So, I have to think like, "Where's the rebar," and "Where's the lock or the key?" And it's just like silliness, but it's just another thing. Do you know what I mean? It's not life or death,

it's just another thing to go on your Mom list of things to do. It's like having another kid that you don't care about at all.

Later in the interview, when asked whether this division of labor was just understood or had been negotiated at some point, Deborah responded: "Is that negotiated? No." Jill expressed a similar sense that mothers were naturally suited to the labor because of their role as family managers:

> I felt like I had a lot of the burden, right, like I had to figure—I was dealing with insurance, I was the one that was dealing with the contractor—I was dealing with the elevation. I truly couldn't do one more thing. I mean, isn't every mom the manager of the things? Yeah.

This sense that mothers were naturally suited to the cognitive labor associated with recovery came with a common corollary that husbands were not up to the task. Even when husbands did assist with the recovery process, the mothers still seemed to be managing all of the cognitive labor associated with it and just delegating tasks to their husbands to complete. Meghan described her marital dynamic this way:

> I am the bookkeeper. I'm the kind of bill-financial kind of person too. So I'm dealing with all that paperwork. But he's really good at like, you know, "Go down to the city and get this permit done," or whatever. Like if I finally get that paperwork from the contractor, then he'll come get it and he'll take it to [our bank]. You know he'll get that done. So . . . he's a really good support person.

Meghan made it sound like her husband was her assistant, who carried out tasks but only when assigned. She still had to manage the process and decide what needed doing, but she talked about her husband's efforts with a surprising amount of appreciation.

While Samantha's husband worked long hours, which she acknowledged, she also noticed a pattern, which she related with a wry expression: "So whenever we have a lot of stuff going on in our personal life, he seems to get really busy at work. Really busy."

Samantha suspected that her husband shirked his family duties when things got complicated, as they had after the flood. When the women tried to speak up to their husbands about their feelings of resentment, it often only made them angrier. Sarah, who worked full time, was clearly still feeling intense rage about all the work she had done, even as we were sitting in her chic new dining room a year after Harvey, the work complete. Sarah generally tried to give her husband the benefit of the doubt, but when she thought about everything she had done, it was hard for her to describe the situation as fair. And when she had complained to him about it, he had responded that he was "letting" her make all the decisions:

> I need somebody to listen to me because I really feel like I do everything, but am I crazy? But it's like—I did all—I mean, I was the one that was responsible for the insurance. I'm the one that found the contractor. I mean, I handled all the [architectural] drawings. I handled all—picking out everything. He was just like, "Well, I'm letting you do whatever you want, I don't care, it's up to you, it's fine." And I'm just like, "It's a full-time job. I mean, it is a—you know, rebuilding a home, that's a, that's a full-time." I, I did it all, and he'll admit that.

According to Sarah, her husband felt he was doing her a favor by not interfering in her design and construction decisions but didn't understand just how much work that meant she was saddled with. Sarah wasn't the only mother to describe managing the recovery process as being like a second full-time job. And because these mothers also bore more of the household labor, for many this was a *third* shift on top of their second shift.[12] Over time, the unequal division of recovery labor bred deep resentment within the marriages, which resulted in multifaceted conflicts beyond the general stresses of the recovery process.

While most of the mothers, despite feeling resentful, kept their heads down and ground out the recovery labor until it was done, believing that to be the most efficient way to get their families back into their homes, a few of the mothers staged protests or even full-scale rebellions. Many of the husbands understood that if they were not the ones doing the recovery labor, they should stay out of it—but a few of the husbands constantly made suggestions that were not well- received. Two mothers grew so irritated with their husbands' repeated interjections that they went on strike. During her initial interview in November, about nine weeks after Harvey, Deborah, who worked full time, told me:

DEBORAH: In any event, for a period of time I was the general contractor of the repairs. I got as far as the cleanout was done. I had spraying done for mold remediation. I had insulation put in the walls and drywall put on the walls. I would have been done by now except my husband had really great questions about "Why are you doing this this way" and "Why are you doing that that way." And I said, "Here's what I'm gonna do." And he said, "Do you think we should consult an interior designer?" And I said "Absolutely, but I have a framing guy coming tomorrow. So if you wanna consult an interior designer I have to throw off my whole schedule. You're gonna have to take over the project." And so we've had zero additional progress since that day. But he is in charge of the project now.

INTERVIEWER: [*Laughing*]

DEBORAH: Your transcriptionist can't see me smiling.

INTERVIEWER: She's smiling.

DEBORAH: My husband is in charge of the project now.

INTERVIEWER: And you're good with that?

DEBORAH: No. Everything about it is terrible. I'm still smiling.

Deborah's smile was really more of a grimace, a face that told me, "I knew nothing would happen once he took over." The lack of progress was killing her, but she could stubbornly wait her husband out if he wanted to be in charge. Similarly, after Maria had collected multiple bids from contractors (each of which involved rearranging her work schedule), her husband rejected all of them as too expensive.

Annoyed, Maria told him he was on his own: "So I had contacted a couple, and the bids we got were really high, and, and he wasn't willing to sign on to any of those. And I got frustrated, and was like, 'Okay, fine, then you go find a contractor.' And for three or four months, nothing happened."

The mothers who relinquished control to their husbands seemed to feel vindicated when "nothing happened," as though they had been proved right about their husbands' shortcomings when it came to project management. A few of the husbands were, in fact, eager to dig in and work on the renovation, but this was not always welcomed by their wives. One mother, a repeat flooder, memorably shut down her husband's initial idea of handling the renovation himself:

> When we flooded the first thing he said to me was, "We don't need a contractor. We [just] did this last summer. Like, we can do this ourselves." And I said, "Okay. When you go out, dig a hole in the backyard that's approximately six feet deep by about six feet long and two feet wide, just so there's a place to dump your body."

Some mothers reported little postflood conflict over the division of labor, but this seemed to be because they just did all the work themselves. As Cathy put it, with two rather contradictory ideas in the same sentence: "Yes, I think [the flood] made our marriage stronger. I mean, he was gone and so I just had to do it all." At first, Cathy thought the year had gone well in terms of conflict, although that was probably because she was managing the recovery alone. Julia also experienced little conflict in her marriage, since she was managing most of the recovery aspects. She reflected upon her year of extra labor after the flood and the result of her efforts to "shelter" her husband from the stress and the work:

> He does not understand that, because it's like, we shelter our kids from the bad so—to tell them "It's okay, it's okay, it's okay," you know, because you don't want the kids to be traumatized. I think maybe I did that too much for my husband and didn't traumatize him enough.

And now he's like; it's—"If we flood again, we'll just—we'll do it again." [And I'm like,] "I don't think you understand."

For Julia, not engaging her husband in the day-to-day contractor drama and project management details meant that he believed the flooding was a pain but not that big of a deal. The family could just renovate again if the house flooded again. She felt that her labor and stress were hidden from view—by design—but now saw the downsides of this strategy. Maybe a little more conflict over the division of labor would have been a good thing. Like many of the other mothers, her resentment and frustration about the situation were clear during our interview—but she also clearly saw her own reactions and behaviors as complicit in the situation she now found herself in.

The resentment over the division of labor after the storm and the more general stress of the event had serious repercussions for some of the mothers' marital relationships. Mothers described three main types of romantic relationship impacts in the year following Harvey. First, the mothers admitted that their own irritability was high, which led to increased day-to-day conflict with their husbands, something typical after disasters.[13] Next, the stress of the recovery period caused many couples to neglect each other in ways they now regretted. Finally, the crisis forced some of the couples to confront long-standing issues that threatened the stability of their marriages. Overall, the mothers believed the year's stresses had created a crucible for marital conflict, one with all the components to put their relationships under threat. As one mother described it, the year was freighted with emotional weight:

> So, it's definitely taken, I think—there's been an underlying just emotional kind of aspect to the year—and—you know, it's been stable, but there's just kind of this underlying, you know, discomfort because you're not quite sure what you're doing—and new information kind of comes up every couple months or every month, and so—and you're trying to figure out—weigh that in the midst of, you know, either doing jobs, or taking care of kids.

Rebecca, like many of the mothers, found her husband was a convenient person to direct her anger at over the entire situation. As the family logistics manager who prided herself on her ability to figure things out, the fact that she could not solve every problem was extremely frustrating:

> So, I definitely am more irritable—and, and again, I think that all this came from frustration of like, "Why, why isn't this—like why isn't this better? Why is all my hard work, and my smart brain, and—why, why is this not better? Why, why does everything have to suck?" Well, I'm working so hard, and trying so hard, and stuff, and, and maybe it's sucky because of [my husband].

For Rebecca, her resentment over the division of labor boiled over and impacted Paul; after all, Rebecca was giving it her all. If things weren't working well, it had to be Paul's fault.

Mothers who reported little substantial conflict did report noticing other types of relationship impacts, specifically on emotional intimacy and sex. For both the mothers and their husbands, the intense stress of the recovery year meant that nurturing their marriage was at the bottom of their to-do list. A year after Harvey, this was something they were beginning to pay more attention to. Kelly put it this way:

> And our—like our relationship is—like my husband's and mine's relationship is still strong, but it's—like the other day, we went on a date, and we're like, "When did we—when have we done this?" Like I think it has literally been probably a year since we've gone and done anything that we would call dating. Like both of us really buckled down, and we just had so much to do that we just sort of neglected each other. We didn't necessarily fight, but we just—were just like in parallel tracks.

Many of the mothers described feeling like they were in "parallel tracks" with their husbands during the recovery year: all the labor the mothers did took away from their leisure time and time they might otherwise have spent with their husbands.

In their temporary homes, often small apartments with their children closer than usual, or living with relatives, mothers experienced what was for many an unusual lack of privacy. This lack of privacy, combined with the stress of relocation and renovation, as well as the strong undercurrent of resentment around the division of labor postflood, meant a collective lack of sex. As one mother put it, given everything she was managing, "That is the last thing I have interest in. I don't want to touch him, I don't even want to look at him." One of the youngest mothers described the discussion in the neighborhood social circle:

> Oh, it's been hard. Yeah, so that is an interesting issue. It's just anytime there's so much stress, you know—I'm living with my parents, but he's living with his in-laws. Yeah, we—I was joking with some other moms one time, like, "The entering kindergarten class of Bayou Oaks in five years is going to be very light."

There weren't a lot of babies being conceived in Bayou Oaks in 2018.

Not all of the mothers experienced increased relationship conflict during the recovery year. Some felt their relationships had improved, or that the test of the flood showed how strong their relationships were. Laura, for instance, saw her husband as a partner: "And I was like; well, no, actually, that—this is horrible, but if I've got to do this horrible thing, I'm glad I'm doing it with him." Similarly, Mary said of her relationship with her husband over the past year:

> So, I think, I think we're good. I mean, I think we're—we've, we've been good, you know. We have, you know, our moments where we're, you know—and the night of the flood, you know, was terrible. You know, it was not a shining moment, but overall, we've been good. I think it's strengthened our marriage. I would say—I mean, it's—I would say it's a—been a crappy year, but it's been a crappy year together. Does that make sense?

When I first spoke with the mothers, shortly after the flood, most reported some low-level conflict around the early decisions they had to make, but they expected things to get back to normal with their relationships once they were back in their renovated homes or settled into a rental home while they lifted up or rebuilt the old ones. The follow-up interviews revealed, to their surprise, that rather than returning to an equilibrium, some of the marital conflict did not improve when the families were back in their homes:

> It was one of those things that, you know, twice a month there was something that I would completely lose my mind on him about . . . just completely. And just . . . I mean, just the stress of everything at that point. And so even when we'd moved back into the house. . . . Yeah, I was being resentful and awful when we were actually back in the house. 'Cause everything should go back to normal, and it really didn't.

The mothers had so romanticized being back in their homes that they had anticipated a fantasy, stress-free family life that often failed to materialize. When nothing mattered except getting back into their homes, it was easy to believe that walking over the threshold would result in the longed-for family harmony. But moving back in was not the equivalent of winding back time to before Harvey hit. Each family had endured trauma, stress, and financial strain in their time away, and that left fractures in their family lives.

Many of the mothers reported additional conflict at home, whether related to the recovery or to ordinary marital strain, but for some those conflicts were serious enough to threaten the relationship itself. However, the mothers reporting major conflict did not see it as something that Harvey had caused; rather, they believed the flood had exacerbated existing cracks in their marriages. One mother, for example, had long felt that the family needed more income. The additional expenses brought by Harvey had made clear to her that she needed to finally push her husband to work longer hours. She also feared that her husband would resist and that this conversation might be the final straw for their marriage:

> I knew who I married, you know, and I, and I love him to death, but there are certain things at this point in time that, that he's simply going to have to—as uncomfortable as it's going to be—he's going to

have step up. And as uncomfortable as it's going to be for me, I'm going to have to—you know, once I start going down a certain road, it may not go the way I want it to go, so I've—not neglected, I'm just reluctant to do so because I don't want—I want to stay married. I don't want to get divorced.

The flood surfaced a relationship problem that this mother would have otherwise continued to ignore or postponed discussing.

Another mother, a repeat flooder, described her marital stress this way:

The past year has really been so wretched. Really bad for us, really bad for the family. I—you know, a couple times wondered how he and I were going to make it, and then I realized that we didn't have enough money to get divorced, so we had to stay married. But it was—I mean, the stress put us on really opposite sides as far as how we managed it. And was super alienating and just messy, and the kids felt it, and we saw it in their behavior. And it was bad, bad, bad, and I have now got everyone in therapy including us—so I feel like—I feel like there's light at the end of the tunnel. So, I'm feeling better, and everyone's showing healthier signs, but it did an ugly number on us—It did a lot of damage. It really—had we been able to afford the divorce, I think it would've happened. I mean, it was that bad.

This mother had been through flooding more than once, and the experience had badly injured her marriage. She believed the only reason they were still together was financial.

For Tara, a full-time working mother of three, the frustration over the division of labor in her relationship came to a head in the year following Harvey. When we sat down in her office at a local nonprofit for our interview, she barely waited for me to turn on my voice recorder before she told me she had divorced her husband. At first I was truly surprised; I couldn't remember anything significant from the first interview pointing in this direction. Later, with the benefit of hindsight, I re-read our first interview and realized that the signs were there. He hadn't helped much with the preparations for Harvey, and Tara was doing a lot of the logistical labor afterward. The added stresses of Harvey and its aftermath laid bare for Tara just how problematic the relationship had become. She was not sad as she described the end of her marriage, though. She seemed lighter; freer, as though she couldn't wait to tell me about it, describing how she felt that God had spoken to her:

> It was like this is—"You made the right choice"—because, for me, it was—I was carrying around a 210-pound weight, you know. I was doing everything, and that's what I realized. So, I was like; I'm doing this all on my own anyways, right, and now I'm doing extra, and I'm parenting another person that needs to be parenting themselves.

While Tara was the only mother to divorce in the year following Harvey, three others reported that they had contemplated it but were deterred by the financial implications and by the prospect of further disruption for their children. And even the mothers who didn't contemplate divorce described intense relationship conflict during the year, mostly driven by resentment over the division of recovery labor. In addition to this significant stress in their marital relationships, the mothers were burdened with the considerable emotional labor of making sure their children were all right after experiencing flood trauma. Just as the home recovery process fell into their domains, so, too, did watching and caring for their children's postflood reactions and behaviors.

The mothers' postflood narratives related to their children were largely full of the ways they had endeavored to protect them from the worst effects of flood trauma, displacement, marital conflict, and recovery, something universal after disasters in both high- and lower-income communities, but in the weeks and months after Harvey, cracks began to show in the veneer of maternal protection.[14] Several weeks after Harvey, Cathy was driving her children and a neighborhood friend's daughter home in pounding rain when she got a flat tire. Driving in the rain by itself was enough to, as Cathy put it, "test all our PTSD." As the rain poured down, Cathy pulled over and tried to figure out what to do. Meanwhile, the kids in the car were freaking out:

I had one of [my son's] friends in the car 'cause I'd taken them back from tennis and was taking her home. And she'd already flooded twice. And she was flipping out. And they were just . . . all three of 'em were crying. . . . I had a flat tire and I made it all the way back to the apartment. Not the best idea, but once again I was like I don't even know who I'd call right now. There's no one in the neighborhood. It wasn't like I could just go to a neighbor's house. Like [Bayou Oaks] was vacant. It was raining, and lightning, and thundering. And like the little girl kept saying, "Is this another hurricane? Please tell me if it's another hurricane." And then at one point she says, "If this is another hurricane, that's it! I'm done!" I was thinkin' "Me too, sister." I was like, "I hear ya."

As the guardians of stability for their families, the mothers were eager for signs that their children were "all right" after the flood. Many of them at first told me their children seemed absolutely fine, but gentle probing helped them remember instances of their children obviously processing the trauma. These instances, however, were often downplayed by the mothers as minor, and in some instances they truly seemed to be. Some mothers had seen no signs whatsoever of trauma in their children. For most, however, some indications that the children were working through what had happened to them appeared by the first interview and continued across the recovery year. The reactions varied with the age of the child. The younger children showed their worry in two primary ways: through play and by assessing their own level of safety by repetitively querying their mothers. Older children tended to ask more sophisticated questions about their family's future plans and, accurately assessing their mothers' stress levels, to try to take some of the flood burden on themselves.

Children often express complicated feelings through play, especially after a traumatic experience.[15] One Bayou Oaks mother described her young daughter's "rescue play," something recounted many times in the mothers' interviews:

When she's playing she kind of has a lot of stories about the flood. Like, she started playing with cars or whatever . . . or babies or something. And she's like "Oh, I'm gonna go rescue you because you flooded." You know? Or she ... I think she even once drew a picture of, like, "Oh, this is a home I built for my mommy and daddy because they flooded."

Other mothers shared stories of their children wanting to help fix the house as well. Even Emily's 2-year-old was expressing himself through play:

> Like to this day [he] talks about the house being broken. Like yesterday, he was like, "I'm gonna call the dinosaurs and the dinosaurs are gonna come fix our house." And even at school—'cause he goes to [day care]—they had [play] tools. And one time he was picking up the tools and he was like, "I need ... I need it to fix my house."

Another mother noticed her children playing "trapped" and internalized it as a reflection on the situation she felt responsible for putting them in: "And I didn't realize that my son and my daughters play to this day 'trapped'—like since the flood—that they're stuck somewhere. All the time. Everything is, 'Someone's stuck.' I'm like, oh buddies, I'm sorry." This mother felt she had to apologize to her children for putting them in the position of being flooded and experienced her children's emotional processing as a self-rebuke. If it was her job to keep her children safe from harm, she had not held up her end of the deal. Fathers' responsibilities for children's emotional recovery were never mentioned by the mothers; instead, they seemed to view this as squarely in their domain of parenting responsibilities.

In addition to expression through play, other children expressed themselves more through their language and by repeatedly asking their mothers if places or special items were safe from flooding. This aspect of children's processing was nearly universal in the mothers' stories:

> [My son] has been going back and forth where he goes, "I hate the house!" and "I never want to go back to the house." And he keeps asking if places have upstairs ... for awhile, when we were going to

places. I'd be like "We're goin' to [friend's] house." And he'd be like, "Does it have an upstairs?"

Children were concerned about going over to homes that did not have an upstairs to retreat to, in case of flooding. Leah's young son, during the cleanup phase, insisted on being held while they were inside the house. He didn't want to let his feet touch the floor, because he remembered having to wade through water in the home. Like many of the other children, her son was also cognizant, post-flood, that items that were higher up were less likely to be ruined. Like many of the other children, he used "flood" as a term for something that could happen to an object, like a toy:

> [He] will sometimes like, be playin' with something, and then he'll pick it up and move it, you know, up onto a counter or tabletop and say, you know, "I don't want this to flood." And you know it's kinda heartbreaking. Or if it's raining, they'll say, "Is it gonna flood?" And we'll say, you know, "No, that was really unusual. That does not happen often."

Older children used more complex language and ideas to make sense of the flooding. Although the mothers seemed less worried about their older children, recovery trajectories may be harder to monitor if adolescents internalize their trauma, and effects of disasters can linger for years.[16] Moreover, mothers' and children's trajectories after disaster may diverge in ways that are challenging for mothers to detect.[17] Janet recalls how, after the Memorial Day flood when her family was flooded for the first time, her son developed his own lingo, which recognized how his life was completely different before and after the event: "But you know, there was always 'before the flood.' That was kind of [his] term—it was 'BF' and 'AF' for a while." For Janet's son, the flood was so pivotal in his life that he constantly thought of things as "before" and "after."

Some of the older children seemed more concerned about their mothers than about their homes. Experiencing their children's empathy toward them was extremely difficult for the mothers, who believed it was they who should be protecting and comforting, not their children. Shannon recalled that her younger daughter went away with friends during the first Harvey cleanup day, as it was her birthday, and Shannon wanted her to make the most of what should have been a celebratory day. But while away with her friends, her daughter kept texting Shannon to make sure she was all right. Up until this point in the interview, Shannon had not cried. But remembering her daughter's concern was the hardest part of her flood story to tell: "Well, [my daughter], I think, was very worried. She wanted to be . . . she's a child and she wanted to be away from all this and wanted some entertainment. But she was texting me constantly. 'Mommy, are you okay?' She saw me break down a few times."

The older children were not used to seeing their capable mothers break down. Laura, mother of two, noticed as well that in the first days after the storm her older daughter was going out of her way to take care of her, and like Shannon, was overcome with emotion as she described it:

> She really understood I think much sooner [than my younger son], and was definitely making exceptional efforts to try to be more helpful and you know whatever . . . you know, "What do you need?" And when we were . . . those first couple days at the apartment, it was one evening, and she was like, "Mommy could you just . . . why don't you sit and rest for a minute." You know, "Come sit down." [And] the first morning we were there she's like, "I'll make the pancakes this morning." So she cooked us breakfast and [crying] . . . and she's . . . she's been super tough. And I'm trying to give her a moment to be a kid and say, "It's okay . . . it's okay to have lots of feelings about all of this."

For Laura and Shannon, the tables were not supposed to be turned like this. They should be the ones helping their children cope, not the other way around. They were grateful for the empathy and found it moving, but they also felt guilty about it. They wanted their children to be kids, not to have these adult responsibilities.

In addition to the shorter-term impacts on children the mothers were closely monitoring, they were anticipating some longer-term impacts as well. In *Children of Katrina*, Alice Fothergill and Lori Peek found that even seven years after Katrina, the children were still "actively processing and coping... on a daily basis. This was not something they were 'over.'"[18] Children's distress decreases dramatically over the two years following a traumatic event, but it may linger for years beyond that.[19] While many of the mothers described their children's emotional processing as healthy expressions and did not seem overly concerned, not all of the mothers downplayed their children's reactions to the event. Some were intensely monitoring their children for signs of distress and sought out experts to assist them. Elaine, who had a friend who was a child psychologist, was paying close attention to how her youngest was behaving:

> Then my little one who's six does tend to bring it up a lot. Like "Oh, we used to have that but it washed away." She... she brings it up, like that, a lot. And [my psychologist friend] was just telling me that if they talk about it in that way—and they're kind of play-acting or, you know, talking about, "Oh this washed away," but they don't have any bad behavior with it, then it's probably very therapeutic for them to be talking about it like that. So I'm like "oh, then she's fine." 'Cause she definitely talks about it, but not... she doesn't seem stressed out about it. So it's just, like, something that happened.

Cathy felt her children were showing too many signs of stress and anxiety a few weeks after the storm and sought professional help for them. Their trauma therapist had a dollhouse built with water around it, which Cathy found creepy—but effective. She described how her daughter shook visibly as she approached the dollhouse, but ultimately emerged from the session like a new child, less afraid and less anxious. Cathy didn't talk about this experience very much around the neighborhood, though, because she felt some social pressure to downplay her children's stress and anxiety:

> Well yeah, 'cause you feel like—and like I said, I don't wanna be too dramatic... like telling people we've been to the psychologist. Like [they would] eye roll. You know what I mean. I do feel very grateful

that it helped [my daughter] immediately. Immediately. All the stuff that she talked about, and said, and did. And she apparently reenacted everything. I didn't even know she saw all those things [during the storm]. Like it was amazing. And she actually verbally—you know she'll be four in December—but she told me, like, "Thank you for helping me," one night after the psychologist visit.

Cathy was the only mother to describe social pressure to conform to a narrative of resilience for the children, but it was in fact challenging to draw out the mothers' stories about their children's expressions of trauma. And when they did report their children using play or language to express their worries, the mothers presented it as a healthy way of processing the trauma. In fact, many mothers used the events to construct a narrative of the event teaching resilience for their older children, like Amy:

> And I hope it made an impact on her, though they do still like—we were drivin' to school—and [daughter] was like, "I'm just sad that we're not in our house. And our school flooded. And this got flooded." And I said "Okay." I said, "Me too. But what happens if we don't do anything?" She goes, "Then it doesn't get fixed." "Well, what happens if you just sit there, and like, you can be sad, but like if you don't add an action to that, then nothing changes. So gotta keep moving . . . moving forward with it." She was like, "Okay."

After the Memorial Day flood, Rebecca's then-9-year-old son was less enthusiastic than she expected about her attempts to engage him in conversation about the renovation of his bedroom:

> And I said, "What's the problem, it's gonna be nicer." And he goes, "Mom, I just never . . . until the flood I just never knew that you could go to sleep and everything could be okay. And you could wake up and everything could be ruined." And so I was so glad I was driving a car. Just . . . he would have seen it on my face, and he's very tuned in to me. And so I just kinda took a breath and I said, "Listen [son], the world is going to change on you. And some changes are gonna be bad—like the flood feels—and some are gonna be good, like our house is gonna be at the end. But it is gonna change. And the only thing we can do is

make sure that we're malleable enough to change with it." But it was one of those things that, as a mom, it was just this kick in the gut, because it's like—you talk about loss of innocence—it's like there it was like in one statement.

This time, Rebecca was determined to put her son in therapy, to help him process this second flood, even though he declared he didn't need it. A few other mothers also reported taking their children to see a therapist, and a few went as far as getting a referral or two, but the majority did not investigate counseling for their children. Although initially this was surprising to me, given all that the mothers were juggling after the storm and the financial and relationship stress they were undergoing, their logic of taking a "wait and see" approach to whether their children would exhibit lasting issues makes sense. In addition, acknowledging that their children were deeply impacted by the flood would also be an admission that by living in Bayou Oaks—and especially by deciding to stay in Bayou Oaks—they might be placing their children in danger if (when?) it flooded again.

Some of the mothers, especially those who had been flooded multiple times, took the long view, anticipating potential longer-term issues for their children. In particular, three-time flooders Nancy and Ruth expected some impacts as their children grew into young adults. Nancy also put some of the blame on herself for what they had gone through:

> Oh, and I ... you know I'm really good at that [moms putting too much on ourselves]. I don't know. It'll be interesting to see if they come out more resilient, or if they come out more neurotic. You know? We won't know for a long time what the effect is gonna be. But it's not a passing phase. It's defined their childhood.

Ruth, similarly, promised to remain ever vigilant: "I just know ... I know that trauma can impact people. And so I think it's a wildcard, and I think I'll be watching the hell out of my children as they grow to see where that presents itself."

Just as the mothers viewed emotional caretaking for their children as part of their domain, Ruth saw continued monitoring of her children over time as within her domain. This was a responsibility, she and Nancy felt, that would stay with them for a long time. Their children had experienced multiple traumas on their watch as mothers—and so it was their job to assess and monitor the long-term impacts.

While the mothers monitored their children closely for signs of residual stress and anxiety during the recovery year, prioritizing their children's well-being, they paid much less attention to their own well-being.[20] Everything else came first: the recovery process, their jobs, their children, their husbands, and bringing up the rear, themselves. Nearly all of the mothers reported some kind of stress-related mental or physical health issue they attributed to Harvey. Because during our first interviews the mothers were often still fueled by adrenaline, firmly in "get it done" mode, many of these health issues did not manifest until the follow-up interview. But a year of high stress, financial anxiety, marital conflict, and concern about their children took its toll. The mothers reported problems with joints and muscles, headaches, chronic condition flare-ups, trouble sleeping, intense anxiety, and especially, weight gain.

Leah perched on a white couch in her parents' living room during our follow-up interview, her bare feet tucked under her. As I had seen with many of the mothers, the mental load of the last year was almost visible as we talked—like a force weighing down her shoulders. Although she appreciated the ability to stay with her parents during the rebuild of her home, it did not lessen the stress she experienced (and sometimes exacerbated it): "Mentally, I feel like total stressball anxiety. I think I'm generally a more stress-anxious person in general, but especially right now. Like with a lot of the house stuff, like it just—it feels like too much, like there's too many things." Nancy felt a similar sense of being overwhelmed by everything that needed doing, something she believed was compounded by shock:

So I guess I'm in my own sense of shock. I don't know. I have like holes in my brain. I mean I really... I hate to use flood analogies—but I do not have both oars in the water on a regular basis. I mean I really feel... I am so distracted. My focus is gone. But you know I gotta go and pick out granite and see what colors it's gonna be in. Coordinate when the sheetrock people are coming in. I have to pick up more Swiffers. And there's so much stuff that it's always on my mind. It's really distracting. And I don't feel like I'm as present as I need to be.

The feeling of being overwhelmed was described again and again by the mothers, and some of them termed the feeling "flood brain" or "being in a fog." The mothers also described symptoms of probable depression, from lingering sadness to difficulty sleeping. As Jennifer put it: "I mean like a lot of times I just wanna go home and just curl up in bed. I think it's just you know just... I think it's still kinda raw. So hopefully we'll eventually get out of that. We have to somehow pick up... pick ourselves up." Melissa, fortuitously, had started a prescription for Zoloft, a depression medication, about two weeks before the storm hit, and she believed it had really helped her during the recovery period:

So I was probably using Zoloft like two weeks before the flood and I am, like, [laughing], thank God! I told [my husband], I was like, I think I'm not feeling things as deeply, because I guarantee that if I wasn't on this, I would be crying all the time. And just like, not sleeping at night.

Nicole had a lot of trouble sleeping at night during the recovery year, which caused difficulty throughout her workday and in her home life:

Well it is because I actually... my... my dealing with things is not sleeping at night, but sleeping in the day. So when I get anxious and get, you know, over-stressed, I fall asleep in the day. And so I... I really have a hard time making it through the day. So I'm exhausted at 2:00 p.m. Like, I'm like driving like this [pantomimes sleeping while driving].

As Julia put it, the flood sapped her usual energy, drive, and optimism: "It just, it just takes all that from you and just turns you into like, this get through the day, get through the day, there's nothing good."

Most of the mothers reported weight gain over the recovery year, which they attributed not directly to stress but rather to changes in their usual physical activity and diet. Working out was not top of mind during the recovery year, and especially while they were living in their temporary homes, they were not eating a lot of home-cooked or nutritious meals. Takeout and frozen meals were common, sometimes due to time constraints, including increased commutes home from work and school, but mostly due to not having access to a proper kitchen for long periods of time. As Kelly put it: "My activity level changed so much, and my diet changed, and my—everything changed. Like all my routines changed, right?" Like other multiple-flooded mothers, Jennifer noticed a compounding effect as the floods accumulated:

> I think I just gained a lot of weight. I, I mean, seriously, it's like the first flood, I gained 10 pounds, and before I could lose it, the second flood came, and I think I gained 15 pounds. I think it's time that— after the flood, you just kind of—like, I usually would try to be cautious of what I eat. I stay away from refined carbs like sugar, or pasta, or rice, or bread, stuff like that. And, and it kind of curbs my sugar crave, but when—right after the flood, everything is gone to the wind it's become my comfort.

Although the mothers were aware of multiple health issues they attributed to their Harvey experience, including both mental and physical impacts, they were optimistic that once their lives got back to normal, so would their health. But the experiences of families after Katrina and of the repeat flooders in Bayou Oaks suggest otherwise.[21] When I reinterviewed Jennifer, a twice-flooded mother, a year after Harvey in the common space of her apartment building, she and her husband were still trying to figure out what they wanted to do with their ruined home. Upbeat music played softly in the background as we sat near windows looking out onto the palatial pool. Jennifer hugged her elbows as she spoke, looking smaller, younger, somehow, than when we'd first met a year ago. Her knee moved constantly as she told me her storm anxiety was playing a role in their decision and impacting her life:

> I don't know if I can move back in like the last time without the house being raised [up]. And still having, you know, the fear that it might get flooded again. And up to this day I feel, sometimes, nightmares. Just two nights ago I dreamt of the house being flooded again. So it kinda stays with you. And then before the Harvey flood—just maybe a couple weeks before the Harvey flood—I dreamt about the house getting flooded again. That was before the Harvey... before they announced there was a Harvey. So I'm just... you know it just stays with you.

While the mothers spent most of their time dealing with recovery work and worrying about their children, they didn't have a lot of time to focus on their own well-being. The year following the flood caused significant stresses for the mothers, largely due to the uneven distribution of recovery labor between the couples. The mothers saw restoring the home and caring for their children's emotional needs as firmly within their domain, but the consequences of taking on that much responsibility would linger.

6 To Stay or Go

DOES ANYONE THINK THIS IS CRAZY?

> One of the crucial jobs of a culture is to edit reality in such a way . . . that its perils are at least partly masked.
>
> —Kai Erikson, *Everything in Its Path* (1976, p. 240)

Several weeks after the Memorial Day flood of 2015, Rebecca stopped by her house to assess the progress of reconstruction. Most of the house was still stripped to the studs, but new walls were starting to go up. As she stood in the entryway and watched the workers measuring and fitting sheetrock, she suddenly started to shake. She felt sick, and like something was constricting her chest. She stumbled outside for fresh air, where she was confronted by her neighbor's green construction fence; behind it, the home was being razed so something new and higher could be built. Across the street, another neighbor was doing the same. Rebecca's panic attack was triggered by the progress on her own house, still on ground level: "It was starting to look like a house again." Instead of reveling in the sight, a sign that her family would be moving back in soon, Rebecca felt the weight of her decision:

> And I'm standing there and I'm realizing we're gonna have to move back in this house, like this is what we're doing. And yet I'm seeing all

over the neighborhood all the houses that are being gutted. And I'm thinking, "Am I... am I bringing my children back to a dangerous situation?" Like, "What am I doing?" There's me going, "What am I... like what kind of parent am I? And why do I not have more resources to have these choices that everybody else seems to have? Am I ruining my kids' life by bringing them back into this house?"

Rebecca proved prescient about the likelihood of future flood disruption. After Harvey brought them a second flood, the family's Christmas cards struck a poignant but humorous note: her two children, in coordinating blue outfits, looked at the camera ruefully, wearing paper sailors' hats, her son's pants rolled up to his knees. Behind them stood a chalkboard scrawled with rolling waves.

Like several other mothers I spoke with, Rebecca felt that her family had no other option but to take the flood insurance payout and renovate. They could not put together the $200,000 it would take to raise up the house, and she did not believe they would ever get that money back. Rebuilding from the ground up was even further out of reach, promising to add a half million dollars to their existing debt. If they moved, they would have to sell their primary asset for less than half what they had paid for it. But staying put was not done without trepidation. Like most of the other sixteen mothers in the study who stayed on ground level, Rebecca dreaded the next flood, and as a repeat flooder she was sure it would come.

At the same time, she *wanted* to stay. Her younger child was just starting school, and they'd spent a lot of money to buy into the Bayou Oaks Elementary zone. Her older child had many friends in the neighborhood, and they were also zoned to "the good high school," which he would attend in a few years. This was where she had chosen to be and where she wanted to stay, in large part so her children's lives would have less upheaval given what they'd been through in the last three years:

> I think it's also continuity in a way that I don't want the kids to feel—you know, because our house had some upheaval, and we have had to

move into the apartment, move back [home], move back in the apartment, move back [home], I want there to be some continuity.

Rebecca's emphasis on "continuity" for her children reveals the inherent contradiction that many of the mothers of Bayou Oaks expressed. On the one hand, staying in the neighborhood would surely expose their families to the risk of flooding again. On the other hand, uprooting their families and moving would expose their children to other types of risks, namely changing schools and having to make new friends in a new community. Sociologist Kathleen Tierney has argued that disasters are "socially produced" rather than natural events and advocates a "fuller understanding of the role that social, political, economic, and cultural factors play in making events disastrous."[1] So too, as the Bayou Oaks experience shows us, are responses to disasters socially constructed. From the outside, the decisions the Bayou Oaks mothers made in the wake of the flooding would not make sense without understanding the forces that held them to their neighborhood.

The mothers who opted to stay in Bayou Oaks undertook extraordinary measures in order to do so. They considered a wide range of factors before selecting Bayou Oaks and considered its choice a key part of their curation of family life. Most of the eight mothers who left did so reluctantly, and they maintained social ties with their former neighbors. Those who stayed articulated one or more of four primary "accounts" about the choice to stay. An account is a narrative constructed to explain behavior that people believe may be evaluated or judged by others.[2] Family sociology has a rich history of studying women's accounts of motherhood, work, marriage, and the gendered division of labor, arenas in which, with good reason, women expect to be judged.[3] Sarah Damaske argues that "accounts can be understood as the product of the negotiation between actions taken and the cultural meanings attached to those actions."[4] Staying in Bayou Oaks, the mothers in my study understood, was a decision that would need justification to the outside world, and they were ready with their accounts, which were deeply tied to their ideas about motherhood, and especially of how to be a "good mother."

First, some mothers said that staying was the only financially responsible choice; they had no other financial option but to stay, by which they meant they were not willing to take the financial hit selling their house as-is would incur. Second, some mothers said, the entire city was vulnerable to flooding, so there was nowhere else that would be safer. Next, every mother cited either the strong Bayou Oaks community or its schools as the primary reason they stayed. And finally, the mothers framed the flood experience as a crucible that had helped to educate their privileged children about adversity, something they believed would be useful for their character development. The mothers had carefully selected this neighborhood as the place to raise their children, and even repeated flooding did not shake their resolve to stay. The community was still strong, in their eyes, or even stronger than before, and they were still zoned to the same schools. Bayou Oaks Elementary was being rebuilt, and it would be better than ever. The city was widening the bayou, so it would have more drainage capacity and thus it would take even more catastrophic rains to flood the area again. The neighborhood, they believed, had so much going for it that surely it would rebound. Just as the Bayou Oaks mothers had chosen the neighborhood and school initially for their children, they justified the decision to stay as being best for their children. While they worried about future flooding and did not want their children to experience it again, on balance they believed staying in Bayou Oaks was the smartest decision for their families.

Families facing flood damage from Harvey had four basic choices for recovery: to repair the damage and move back in, to lift up the home, to tear down the home and rebuild it from the ground up, or to sell the property as-is and move. No matter which option their families chose, the mothers all tended to explain that whatever they'd done was their only financially viable option. Mothers who remodeled and moved back in, mothers who had their homes lifted up, and mothers who chose to rebuild all tended to explain that financially, whatever they chose to do was really their only option. Their accounts emphasized the financial soundness of their decision, or at least that whatever they chose to do was financially their *least worst* option. Most of the mothers who chose to repair the damage and move back in were first-time flooders. They tended to believe that Harvey had been a fluke, and that since they had avoided damage in the prior two flood events in the neighborhood, they were protected from all but the most catastrophic events. A disaster on the scale of Harvey, they believed, was unlikely to happen again. So they felt the risk was acceptable. None of the mothers who had been flooded for the first time during Harvey opted to move out of Bayou Oaks. The accounts they offered were largely financial. They did not feel it was necessary to lift up their homes, or rebuild something new, because they believed their risk of future flooding was low. Therefore, it did not make sense for them to invest hundreds of thousands of dollars when they could do their renovations using just the money from their flood insurance. None of them were eager to sell at a loss and move out of a neighborhood they liked. They believed that renovation was the only option that made financial sense.

The mothers who had their homes lifted up usually did so because it was their second or third flood, and they could not justify simply renovating and moving back in again. Most of them also did not want to contemplate moving and viewed the home elevation as their best option to stay in the neighborhood while protecting their families from the risk of future flooding. That the homeowners in Bayou Oaks had a choice at all about how to mitigate their flood risk is interesting. Technically, according to the taxpayer-subsidized National Flood Insurance Program (NFIP), if a home was "substantially damaged," meaning the repairs would cost at least 50 percent of the home's value, homeowners would be *forced* to demolish or elevate their homes. A *Houston Chronicle* investigation following Harvey found widespread manipulation of the 50 percent cutoff in order to allow homeowners to evade this requirement. As the *Chronicle* noted, "Telling traumatized flood victims that they will have to undertake expensive home elevation projects is politically and emotionally difficult, so officials lowball the damage estimates, putting people and homes back in vulnerable places."[5] While officials may not want to tell a homeowner they have to tear down or elevate, most homeowners don't want to hear that either. Some of the mothers with the most damage mentioned pushing their adjusters to assess so they would come in under the 50 percent threshold, but most just prayed their homes would come in just under the threshold. Melissa, after two floods, wanted to mitigate the risk for her family but stay in the neighborhood. Her family, after deciding to elevate, was hoping to (and did) receive the substantially damaged designation, because that would qualify them for a $30,000 NFIP Increased Cost of Compliance (ICC) Coverage grant to help with the cost of elevation. The ICC grant was meant to incentivize property owners, especially those with repetitive losses, to mitigate for future floods. The family had also considered demolishing the house and building something new, but Melissa thought that this would not only take too long, it would effectively price the family out of their own home:

> [If we rebuilt], we would be out of our home for more than a year and a half because of how long it takes to build. So that takes a long time, much longer that we'd be displaced. And I started thinking, "Oh my gosh, I'd be sitting on a million-dollar land"... because I looked up... like one of our friends who did that... who built. I looked at their appraisals or whatever. And it's like, they're all a million dollars. And I was like I cannot afford.... I mean we can barely afford.... We cannot pay [the property tax] on a million dollar property.

Melissa felt that elevating their existing home was the only financially responsible option. It would cost them almost $250,000, but in the end, it would not increase the property's value to a level they could not afford. Texas has no state income tax, so property taxes are high—usually around 2 percent of the home's appraised value. At the time, the families who decided to lift up their houses had no idea how much it would increase the home's value, although they thought that the amount was not likely to fully offset the cost of the procedure. Melissa's experience would seem to confirm that belief, at least in the short run. In 2017, before Harvey, Melissa's home was valued at around $570,000. After Harvey, in 2018, the home was valued at around $265,000 (essentially lot value). In 2019, after the family had invested $250,000 in the lift, Melissa's home was valued at about $480,000. This was still nearly $100,000 less than the house's peak value, pre-Harvey. But the elevation (and accompanying renovation) had recouped a significant amount of value while still leaving the family better off with respect to property taxes. All in all, Melissa was pleased with their decision—except that now they had exhausted most of their assets to finance the lift. She wanted to stay in Bayou Oaks, but it came at the cost of "trapping" the family financially.

Mothers who decided to rebuild from the ground up also felt financially trapped, because of the large increase in debt they had to take on. Emily put it this way:

> Yes, it is way more than we ever wanted to spend on our home, and yes, we're going to be what people call house-poor, but in the end, it's the best option for our family, and we will have this home forever, and ever, and ever, and ever. Like we will never leave. And so, I think once that was decided, [my husband] was—I, I told him, I was like, "Once we sign this, like we are going down this path and I want you to know that this is it." And he's like, "Yeah, I get it." To be able to—like the reason we bought the house where we did is to be able to walk the kids to school, to be able to have neighbors that have kids. Like to be able to say, "Hey, yeah, run over to your friend's house and go play."

Emily felt that she and Jim would never leave. They were sinking so much money into this new house that it would be too difficult to recoup that investment anytime soon. But that was all right, she felt, because Bayou Oaks was a place they'd chosen purposefully. It had the right characteristics for their family. They might be house-poor, but as long as they stayed put, they would have everything else they wanted. Melissa and her husband John felt similarly about their decision:

> We love the house we live in now in Bayou Oaks, and it was like, that's going to be our house for the rest of our life. Like, I have no intention of packing up and moving. So we are going to be in it for the long haul, which is again another reason why we decided to lift and remodel. We're not thinking about resale value necessarily at all. We're just thinking, "What do we want to do?"

Like Emily and Jim, Melissa and John planned to stay put. For them, that made additional financial investment worth it, even if it also meant they were now "trapped" in Bayou Oaks.

The second account of mothers who decided to stay in Bayou Oaks was that "everywhere floods." If the risk of flooding in Houston is high no matter where you go, they reasoned, why pick up and move to a different neighborhood? It was true that many of the neighborhoods near Bayou Oaks experienced flooding during Harvey, and that some suburban neighborhoods had been flooded completely unexpectedly after a dam overtopped in far West Houston. But in fact, large swaths of Houston have never experienced home flooding, even during extreme events. When the mothers argued that "everywhere floods," what they meant was that the nearby neighborhoods with good schools that they found acceptable flooded too. One mother explained her rationale this way:

> Lots of friends were flooded, in [other nearby neighborhoods]. So like, we don't wanna go to any of those neighborhoods. I'd really like to stay with the same school, but okay... fine, let's say we go to—I don't know—another school. We would hate to do that to our kids, but okay, even if we said we would do that, where? You know? What's a good neighborhood with a good school? [Nearby neighborhood] flooded. So you know, when you think about it, like, practically, I don't know. There are pockets within each neighborhood that don't flood, but you have to be lucky enough to find a home in that pocket... to find a good home that meets all your needs.

Catastrophe, they felt, could happen anywhere, so there was nothing to be gained by moving. As Amy put it:

> This one is different, because so much of the city was affected. So like, where? Where do you think . . . like where are you gonna go? For this kind of storm. It's not true for Bayou Oaks as a whole [during a regular storm], but for this kind of storm, like give me a break. All of those [people in the suburbs] were freaked out. So, like, gimme a break.

The tension in this account, of course, is that no other neighborhood in Houston had been flooded for the third time in three years during Harvey. The mothers wanted to implicate nearby neighborhoods as also having flooding problems while downplaying their own risk, because that helped to justify the decision to stay.

Some of the mothers referenced climate change to bolster their account. Nowhere near Houston was safe, in their eyes, and increasingly, nowhere in the world would be safe. If this is your point of view, then it really does not make sense to uproot your family just to go somewhere else that faces the same risk. Melissa was one of several mothers who didn't think there was anywhere that doesn't flood or would not do so in the future: "And it just is almost like the—like if you start looking at; okay, let's, let's move someplace else that doesn't flood. Okay, where are you going to go?"

The justification that "everywhere floods" was so ubiquitous among the group that it seemed clear it was something the neighborhood was discussing. Some Bayou Oaks families saw it as karmic justice, perhaps, after Harvey, when nearby neighborhoods had flooded just as Bayou Oaks had done twice before. That former neighbors who had left Bayou Oaks were being hit again in their new locations helped to rationalize several families' decision to stay put. The one group of mothers who did not offer this account were those who had been flooded for the third time during Harvey. These mothers believed there was something unique about Bayou Oaks's vulnerability, even in flood-prone Houston, and all of them either lifted up their homes or moved away.

To understand why Bayou Oaks families did not move en masse, one has to acknowledge the power of social connections and of collective decision-making. Some prior work on disasters and recovery has found that disasters erode community, as homes and institutions are destroyed.[6] Other work, however, has found a "therapeutic community" response, whereby communities come together to help each other recover.[7] In these cases, communities also build a narrative of why staying makes sense.[8] Most of the mothers in Bayou Oaks described a therapeutic community response to the floods, one that *intensified*, rather than eroded, after each flood event. Now, they believed, they were bonded together tightly, and this "communal narrative" meant that it was hard for many of them to imagine leaving the community.[9]

While some mothers said their decision to stay was based on family finances, and others argued that nowhere in Houston was safe from future flooding, the primary account used to justify staying involved the Bayou Oaks community and its schools, especially Bayou Oaks Elementary. In this massive urban school district, as we have seen, the mothers believed that a neighborhood zoned to good schools all the way from elementary grades through high school was rare. Many of them had purchased their homes in this section of Bayou Oaks specifically for the elementary school; the fact that these areas were also zoned to the "good" high school made the neighborhood a long-term investment in their children's public schooling. The mothers, like other upper-middle-class parents, were explicit that this consideration drove not only their initial decision to buy in Bayou Oaks, but also their decision to stay.[10] As Maria put it: "It was like, 'Okay, we'll pick our schools, and then based on that, we pick where we're going to live.' And I think more people are kind of realizing that that's how it needs to be done if you don't want to pay for private school." Another mother explained it this way: "We did explore the option of like, okay, moving to another area, but it didn't make a lot of sense. We moved there for the school, you know, and like, looking forward, middle and high schools. It's a good place to be in." First-time flooder Nicole said that even if there were another flood, her family would stay, at least while the kids were in school: "We would stay here, yeah, unless both the kids were in college. [After that], you know, we might move somewhere that didn't need school zoning. I mean, that's really what it is."

After initially moving to the neighborhood for the schools, the mothers also discovered a community with a strong pull. They had been through a lot together, and as in other close-knit communities anchored by an elementary school, whole-family friendships arose amid the playdates and PTO meetings and Little League games. Most of the mothers, like Kelly, were aware that these bonds would be difficult for outsiders to understand:

> Nobody will be able to understand it because you think well, I ... you would think everybody's gonna just flee. And they're not. They're just like, "Oh—I'm gonna stay no matter what," you know? Like we're staying. We're spending hundreds of thousands of dollars, putting money into a 1,500-square-foot house so we can stay in our community. Because you know Houston's a really big place. And I could pick up and I could move to [a suburb]. And I might as well have moved to California. Like because once you move out there, you're in a completely different place. You have to find a different church. It would be a pain. [The commute] would take time away from my family. And [my high school-aged daughter], you know, she's going to high school with people that she went to Bayou Oaks Elementary with. And they've got these strong connections. And the parents have connections. And the ... you know our church ... it's just everything, you know.

Sarah agreed:

> The thing that's crazy, though, about Bayou Oaks is that this neighborhood has always flooded. It's always flooded, and yet people still want to move here because of the schools, and because of the location, and because they have short memories. And I think they think they're making a good investment. At least, we thought we were making a good investment. But once you are invested in your community, and you make those relationships, and you have the people to call and take care of your children if there's an emergency, and you know that

they're riding their bike that the mom at whoever's house they end up at are going to text you, I mean, that's hard to duplicate.

For Jill, a twice-flooded mother, each flood cemented her family's desire to stay, rather than eroding it:

> So that [flooding] issue will still be there, yes, but we also really love our jobs, and our friends, and our community. It would take a ton to move us. We love where we are. We love our community. And you know each time one of these awful, terrible things happens . . . it makes us want to stay more, not less.

Even when mothers expressed considerable anxiety about the future of the neighborhood, to the extent that their own mental health was implicated, they valued the community so highly that it seemed to outweigh the potential downsides of staying. As Jessica framed it:

> I mean I've never been an anxious person . . . ever. I could sleep through anything. And after the Memorial Day flood, I started getting anxiety. And I would go, like, three days . . . three nights without sleeping. So um . . . and it's gotten a lot better. But I. . . . [H]ow do you ever get over it? So I'm . . . I don't know. It's like our life is measured by these two storms right now. Memorial Day . . . Harvey. So but no, we're not gonna move. Our family's here. Our community and we can get . . . we can build our dream home. Hopefully, it will be, you know, safe.

Even though Jessica's flood anxiety was high now, and she wasn't sure she would ever get over it, she wanted to build her dream home in Bayou Oaks and hope for the best.

Even when some of the mothers said they might move in the future, it would be somewhere *within the neighborhood* that was situated on slightly higher ground. As Laura put it:

> We would like to have the security . . . the housing security. Personal safety first. But then also is the real estate investment and . . . and it's a . . . it's a huge chunk of um of our financial portfolio. And you know it's . . . it's fine to spend whatever you spend on your house, as long as

you can sell it for what you bought it, great. And in light of everything that's happened in the past three years, and this—you know—pox on our neighborhood, even though this is only the first [flood for us], how is that going to impact the sale at some point? [Interviewer asks if they sold, where they would go.] That's an excellent question. Possibly somewhere else in Bayou Oaks. Possibly [nearby neighborhood]. Um somewhere . . . close by.

On an early October Saturday morning in 2018, Elaine and I sat on a huge sectional sofa in her redone living room, now with proper walls, and I admired her open kitchen with its huge island and brickwork backsplash. Two of her four children were noisily and joyfully making pancakes in the kitchen while we talked. Her contractor was still working through the final items on the renovation punch list; nonetheless, she was thrilled with her renovated home and reveling in the convenience of being able to now use "actual pots and pans" to cook dinner for the family. Even though she had some reservations about remaining at ground level, Elaine was sure her family would remain in the neighborhood: "So, as much as we love our house, the ultimate goal might be to eventually move up—I mean, six feet up. Not this house, but maybe a new build in the neighborhood because we, we have yet to find a neighborhood we'd rather live in."

Sarah understood how her desire to stay in a flood-prone neighborhood might sound to others:

Oh, I would want to stay in the neighborhood, which is crazy. 'Cause why would you stay in a neighborhood that you know will probably flood again. And even if you, like, move [the house] up, your garage floods. And your cars could flood. But I think you also realize, you know, there's a lot to be said for that sense of community. And having neighbors that genuinely care about you.

For the mothers of Bayou Oaks, the neighborhood represented everything they wanted for themselves and their children. Within its borders they had found not just a safe place to live and a great school for their children, but also a like-minded community of people they liked and trusted. To leave therefore became unthinkable. As Tara put it, "We're stayin' where our people are." Cathy summarized how many of the mothers felt about the prospect of having to move after losing their homes and their children's school: "Like, the community's the last thing you have and now you're gonna lose that." As Melissa put it, "We've, you know, planted like our roots here, and so we don't intend to pick them up and go."

The final account to explain why they stayed was different than the others. It was not explicitly connected to the decision to stay, but rather was connected to their ideas about how to be a good mother. Staying, they reasoned, was also all right because the flood experience had actually been *valuable* for the families. Many of the mothers noted that the experience had character-building aspects for their children. They recognized their children's privileged upbringing, and all the advantages of birth, wealth, education, and (for most of them) race that they enjoyed. The mothers believed that there was a great deal their children could learn from the adversity of their flood experiences. From the flood, they believed their children had learned the value of hard work and of family problem-solving. They had learned resilience. They had seen a community come together to help each other. These were values the mothers wanted their children to have, ones they clearly felt were more difficult to instill in affluent children whose lives were curated by their mothers.[11] Here, then, was a shortcut to those values, something that would set their children apart in life. If anything good was going to come from this experience, this was it.

One of the key lessons the mothers believed their children had learned through their flood experiences was that when the family faced something hard, they would work together to solve the problem, that sometimes in life terrible things happen, but there is always a way through them, especially if the family works together. The mothers believed that hard work would pay off, and they believed in working with others to solve problems. And this was something they hoped their children would carry forward. Some, like Tara, had seen evidence that their children were learning this lesson. Tara's 11-year-old daughter had been a big help during the flood evacuation, and afterward as well:

> After everything, as we're recapping with the kids, and I said, "Is there anything that you wanna say about what's happened?" And my 11-year-old said, "You know, I am grateful for this experience because it shows that our family can work well as a team." Yes! It's like, "Yes, you're right!" And she was like . . . she goes, "Well, we did cry, but no one argued and no one yelled at you. Everyone did their own jobs. They did their part." And then she looked at the little kids—"Well, they didn't really do anything but they were very funny." But she's like, "We worked together as a team and we got to safety, and everyone was okay." And I was like "Yes, that is our lesson from this." Right? A total lesson.

Elaine, whose family of six had been living in their small second story during the renovations and eating dinner every night around their children's tiny art table, hoped the experience had taught her kids the importance of doing your best every day and making the most of your circumstances: "I think, you know, the fact that we can make jokes about it, and just do our best in the situation and deal with it, has been a positive experience for them. That we sit at that stupid little table each night and eat dinner."

Amy believed the experience would be beneficial for her children, ultimately, because they had so many other advantages. And she knew from her parenting research that adversity was something useful for children to experience:

> I read 50-million children blogs. I love Brené Brown . . . all this stuff. Like you're always tryin'. . . . My goal as a parent is to create kids with grit and bounce. And those are hard things to artificially manufacture. It is hard to teach somebody how to overcome adversity if they never have adversity. And we have White, middle-class children, with parents who fight sometimes but are together, and are loving parents. Both sets of their grandparents are together. You know they're . . . for the most part, their life is very stable. And it's hard to teach them how

to deal with failure, how to not let something that's hard get them down, how to overcome those things. And this is going to give them all of those things, if we can maneuver it with grace for them to learn some of that.

Many of the mothers used the language of "resilience" to describe what they hoped their children had learned. Kelly, for example, believed the experience had been an "opportunity" for her kids:

> My mom was here this weekend and we were talking about it. She was like, "How do you think this affected your kids?" And she said that it taught them . . . she thinks that it taught them resilience. And I was like, "You know I think that's right." I think it taught them a resilience, and a work ethic, and an ability to face adversity. And to watch people face adversity and then come out on the other side, and not have it be defining, that they would not have had an opportunity to see otherwise.

Laura also hoped her children would learn resilience from the experience:

> And my hope is that ultimately it just lets them know that they're tougher, and stronger, and more resilient than maybe they thought before. And . . . and the scene I keep trying to emphasize with them is we're in this together . . . the four of us. We're gonna make it work. We're gonna be okay. But we've got to be here to support each other. Hopefully that will also stick with them as they grow up, and become adults, and face whatever challenges that they will inevitably have.

Deborah also frequently consulted parenting information, especially scientific research, while raising her four children. What she found there gave her confidence that this experience would be useful for her children's development:

> [This study I found] gives a trauma score to each child's childhood. And it turns out that getting a zero on the trauma scale is actually bad for children's outcomes. And so I think I look at just a very small amount of trauma is actually helpful to children. And I do think this

is a very small amount of trauma, so it's like okay. I think they're basically okay. They don't have real behavioral changes. They don't have nightmares. My eight-year-old, who's particularly sensitive, had one nightmare. I think they look a lot to their parents to see how they react. And aside from like arguing with each other about the house, I don't think we've been like, "woe-is-us" kind of thing. And I think they've been basically okay. So I think . . . I hope it will be a net positive for them . . . eventually. Is that terrible?

Just as the mothers chose Bayou Oaks for the warm community and excellent public school, with its focus on world languages and cultures, they now viewed their children's flood experience as accumulating another advantage for their children. They believed experiencing some adversity, or a "very small amount of trauma," could *inoculate* their children against future adversity.[12] They would know how to handle it when life threw it at them. Not an experience they would have chosen, of course, but one that by the end of the recovery year, they viewed with grudging acceptance. Two of the mothers took the standard narrative of building resilience and learning the benefit of hard work to the next level. They believed that the flood experience was valuable for their children because now they would have something to write about in their college applications. As one of these mothers put it:

> And I keep telling [my son], you know when you go to college, write about this experience in your application essay 'cause you know, not a lot of people go through two floods in two years. I keep telling him you should be proud of yourself 'cause you've been through a lot. And a lot of people don't have that experience.

Similarly, the other mother thought the experience might set her children apart in college admissions, although maybe not if they stayed in-state:

> Oh, they will put everything down in their college essay. I think partly because their experience is gonna be particularly unique among people that they will . . . the applicant pool that a school will be

looking at. Again, if they're looking at going to [local university], or [flagship state university], it might be a little different. But anyway in my mind I'm thinking they might apply across the board of the United States. And in that case, they would be very unique in their applicant pool, and the experience that they had in this particular event, and what it could say about them, and what happened, and how they came out of it, and how they persevered. All the things that college essays want to know about.

These two mothers, as guardians of stability for their families, were thinking ahead to when their children would apply to college. If they had to go through this terrible experience, at least it could result in something positive for their children's future chances of success. Unlike the rest of the mothers, who framed the experience as something to help their children learn to overcome adversity and to improve their moral character, these two made the lesson explicitly transactional: something their children could use in the future to their own personal advantage. Adversity was hard to come by in an upper-middle-class family, and they believed the flood had produced an unexpected benefit for their children's college application essays and admissions prospects.

The mothers were also interested in how the experience would shape their children's impressions of humanity. While most of the mothers talked about the experience as helpful for forming their children's character and value set, in terms of resilience and seeing their parents work hard to get through something, virtually all of the mothers also believed that their children's seeing people come out of the woodwork to help after the storm was beneficial. This, they believed, would shape how they generally felt about others, and they wanted them to feel positive. As Leah put it:

> I think long-term it might be good. You know they've had all this time with their aunt and uncle, and seeing the family helping family, which I think is something important and I'd want to instill in them. And seeing the community helping each other out. I think overall, the long-term things that they will take with them will be mostly positive.

When Mary and I talked about what good might come of the flood experience and what she hoped her daughter might learn, she teared up. Wearing a "Thankful, Grateful, Blessed" T-shirt, she said:

> So yeah, I will also look back at it as a time where whenever I get disappointed in society, this will be when I look back and go wait, people are good. They really are. When it comes down to it, people are good. And you forget that in the day-to-day.

In the end, the choice of Bayou Oaks had not just situated their children in one of the best neighborhoods in the city for urban, mostly progressive parents. They had a strong community and a great school, and their children had a diverse set of friends at school. The choice had also given their privileged children some adversity, which would help them grow as people and be better adults in the future. Exposing their children to adversity helped to make them *good* mothers: mothers who prepared their children to face future challenges.[13] And the mothers believed that for their children having learned those lessons and been drawn ever tighter into the community in the neighborhood, it would be disadvantageous, cruel, even, to remove them from Bayou Oaks. Their kids had lost so much and experienced so much turmoil. Even if it was a good life lesson in adversity, it was enough. They didn't need any more. Uprooting their families would just add to the strain on their children. They should stay where they were and ride it out. Bayou Oaks was home.

At the tail end of our first interview, six weeks after Harvey, Cathy, a first-time flooder, sat at my dining room table twisting a damp tissue as though she were anxious. We'd been talking for two hours, but something was clearly bothering her as we wrapped up. I watched her, wondering what she was thinking, as I tidied up my notes and turned off the tape recorder. "I feel like I need to tell you something," she said, "although I probably shouldn't." My interest solidly piqued, I asked if this would be off the record or on. "It can be on the record," she said, and I turned the tape recorder back on. While almost every mother (including Cathy) had given me one of the primary accounts of why they were staying—finances, the lack of other options that would not flood, the community, the schools, or the life lessons their children had learned—Cathy wanted to peel back the curtain for me. She loved the community as much as anyone, and the school had been unusually welcoming to her son with special needs. But she also looked at the situation with a slightly different eye:

> So most people are coming back. And um . . . it's a real . . . you feel very uneasy. I don't openly talk about that . . . about those kinds of feelings. Because I think it feels like you're rejecting their plan. Every single person wants to know, "What are you doing? So what are you doing?" That's the first thing . . . even now when I haven't seen someone and talked to them for a long time, like you see 'em and say hello. And it was emotionally. . . . I mean if you haven't seen someone yet, you give each other a hug but you then immediately wanna know, "So what are you guys gonna do? What are you doing? Where are you going? What are you doing? Are you building? Are you raising your house? What are you doing?" I mean that's all of our conversations. So if you're like, "Yeah, I'm leaving," they'll be like, "What?! You're leaving?!" And they're like, "They live on [street that flooded only during Harvey]. They only flooded the first time . . . they're leaving? What?!" Like you just . . . I would never just openly say that. I would never. But everybody wants to know what you're doing. And I think it's because you wanna feel justified in what you're doing. And I get it. I totally get it. But I feel like once again I've seen this [during the prior floods]— and been on the outside—and been able to see it. Now I'm in it. And I'm like, "This seems crazy. Does anyone think it's crazy that we're all just gonna kinda keep flooding? Like, what are we doing?"

Mostly unspoken during all of the interviews was the sense that the future of Bayou Oaks depended on other families staying too. So convincing families to stay was in everyone's self-interest. Not just for property value reasons, although that was there, but for the community's future as a whole. If families started to flee, it would hurt everyone who chose to remain. And of course when families left, it undermined the mothers' accounts of staying. So for someone like Cathy, who had flooded "only" once, to "reject" the community's "plan" would result in social censure. In the end, Cathy, like the other eighteen first-time flooded families, decided to stay. But she was decidedly less optimistic about the neighborhood's future and more openly concerned about the impact of future flooding on her family. During our second interview a year later, this time conducted in her small but pristine white dining room that looked straight out of a designer's showroom, she again expressed her uncertainty about her decision:

> It starts to feel like how much more of this should I put my children through? Is this the best idea? I mean and now I'm seeing how long it's taking [my son] to get out from under this. And now that he's gone through it, if there's another flood—and we don't [personally] flood—but again we're in the front row, do I want to keep watching my friends go through this? Like [watching] my community go through this.... But I don't, I don't have a Plan B. I don't know what my alternative is in this city, I don't know what else I would do. We chose that elementary school very much for my son. I don't know where else I would want to put him.... It's so hard because I don't want to leave, but you feel like—well, I mean, just like, probably people in the outside world, are like, "What, what are you doing?"

Even though Cathy realized the need to justify her decision to stay, because outsiders won't understand it, and felt the strong pull of the school and community described by the other mothers, she was staying reluctantly. She did not feel she could talk about her fears with the other mothers in the neighborhood, for fear of seeming like she was abandoning the community. She worried about the future, not just for her own family, but for her neighbors in pockets of the neighborhood with higher flood risk.

Outsiders hearing about a neighborhood that has been flooded three times in three years might well ask why anyone would choose to stay there, and the rationales might be difficult to grasp. For the eight mothers who moved out of Bayou Oaks after Harvey's devastation, leaving was clearly the right thing to do, but they did so reluctantly. They took major financial hits by selling their properties at a loss, but they judged that the peace of mind they believed they would gain was worth it. And they hoped to retain their friendships with the Bayou Oaks residents who were staying. They did not like leaving their neighbors—it felt like they were abandoning them to face future flooding events on their own—and they did not like leaving the school community they loved. Three of the eight families who left had children who were now aged out of Bayou Oaks Elementary, and two other moving families were private school families, so the hold of the neighborhood was less for them than it otherwise might have been. Of the remaining three families, two sent their children to the public school in their new neighborhood, and one continued sending their son to Bayou Oaks Elementary on a magnet transfer.

While the mothers moved reluctantly, they were nonetheless convinced that it was the right choice. They did not equivocate after the fact. They all reported missing their former neighbors, however, and they worried for those neighbors' futures:

> I'm probably the one [in my family] that suffers the most from leaving our block because I had the most—connections. And so, yeah, it's, it's going to flood again, and I hate—you know, and so I've had that conversation. I've said something like that in front of [a former neighbor] and I, I, I could've like slapped myself silly, you know, because you—I know better [than] to say something like that, but I'm like, you know—because I think I said, "When we flood again, I will come get you—know I will come get you." And I'm thinking, he doesn't need to hear that, why did I just say that, you know?

Even as they worried about their former neighbors, they were incredulous at the choice to stay, especially for those neighbors who like them had been flooded twice or more:

> It's totally different, but this was just like too much, and it was not safe. Yeah, so—I mean, I don't even—I didn't really feel comfortable even with the idea of like lifting or building even if we could've

afforded a brand-new house there because it's like building your dream home on shaky ground, you know, soggy ground.

Anna and Deborah were counter examples to this group of reluctant leavers. Both sent their children to private schools and so were not as enmeshed in the Bayou Oaks Elementary–centric networks. For Anna, the neighborhood's attraction was that it was a safe and affordable place to live that was convenient to the medical center. Although she had developed a few friendships with neighbors, her family's primary network was through her children's school and local family members. Anna had grown up in California, and her family frequently traveled back and forth to visit Anna's mother; Anna had long hoped to move back to California permanently. When Anna's mother passed away suddenly about six months after Harvey, Anna inherited her mother's home, including its low property taxes. With the family living in an apartment while they decided what to do with their flooded home, suddenly everything fell into place: they would move to California at the end of the school year. Thus, even if Anna had felt pressure to stay, it would not have mattered.

Deborah was a leader in the local Jewish community, deeply embedded in multiple local institutions. Although she was connected to the local Jewish community, she was not tied to Bayou Oaks per se (however, her husband had grown up there and insisted it was "the" neighborhood for affluent Jews). Deborah was not intimidated by pressure to stay in the neighborhood, and in fact went so far as to make the argument for leaving in local civic meetings and on widely read posts on social media. When I reinterviewed her a year after Harvey, she was living with her family in a rented house in Bayou Oaks while they built a new home on a double lot in a nearby neighborhood. Deborah was excited about the new location, which would be less than a mile from their synagogue, about half as far as their former home had been. When I asked if she felt nervous living in her temporary rental home in Bayou Oaks, she laughed and said, "It's a rental! I truly don't care what happens to this house." Now that she knew her family's future would soon be secured in their new, elevated home, she was happy to rent something in Bayou Oaks while they waited for the house to be completed. Her husband, who had found the rental property, told her that a key feature was that it was "an easy kayak to Kroger." They would, he wryly suggested, be prepared for the next potential flood.

While the mothers who left the neighborhood were convinced of its high risk of flooding again, those who stayed held varying opinions on the matter. Mothers' views about whether Bayou Oaks was likely to flood again were strongly related to how many times their homes had flooded. First timers often said that Harvey was a fluke; because they'd been spared from the first two floods, they believed their homes were more protected from future flooding than others in the neighborhood. This group of mothers was aware that they might be wrong, but they still felt pretty confident about their future flood risk, like Michelle:

> I mean, it's probably really Pollyanna stupid of me, but I, I, I do feel like there's some—I do feel like it's kind of a fluke, like I, I do. And for it to hit again, and, and right here and all, I mean, I just think the likelihood is low, let's put it that way.

Amy thought another flood was possible, but only if there were another massive storm like Harvey:

> I think for us—and this might be because time has passed more so than how I felt in that moment—but I've come to a place where it would take another Harvey for us to flood again. We're in a different category than the repeat flooding people. And if I was a repeat flooder, I would be making very, very different decisions. But as a first-time flooder in a very big storm, I do not have confidence that it'll never flood again because it did. It's happened. So like it's clearly possible. But it would take a storm of this . . . this magnitude, and it would take [the city] doing nothing on the drainage stuff, and both scenarios are entirely possible.

Sarah also felt that it would take another catastrophic event for her house to flood again, in which case:

> And then we're all screwed. I mean, we're all screwed anyway, but I mean, it's like—it's, it's the bigger national story of screwdom. But I do think—at that point, we would probably look to find something that hadn't flooded in, in the neighborhood, or something that was raised that we aesthetically liked.

The multiple flooders, however, had entirely different views on the possibility of future flooding. "They're fucking stupid," said one, who had a good deal of empathy in general for the first timers but not for their future risk assessments. Rebecca felt confident, after having been flooded twice and renovating twice, that she would be flooded again:

> I guess I picture myself being in this house in ten years, but I—when I really think about it, I can't imagine there'll be any houses like mine in ten years, or if they are, they'll be elevated because, again, I don't think there's any part of me that thinks I'm not going to flood again. It's just a matter of time.

Molly, a three timer who elevated her home after Harvey and moved back in, felt pretty confident now that her home was high off the ground and was ready to close the book on the last several years of her life: "It's been a three-year ride of hell. Actually, I guess, we're on four now. Yeah, it's four now—but, hopefully, we're done." Ruth was also feeling confident, with her new fortress of a home currently under construction and ten feet off the ground: "If we flood at ten feet, everybody is going to be . . . it is going to be an apocalyptic death kind of situation." After three floods in three years, the Bayou Oaks mothers had differing opinions about their own personal risk of being flooded again, and these beliefs were tied to how many times their homes had already been flooded.

Even when mothers felt their future risk of being flooded was remote, they usually admitted taking some steps to mitigate future danger and damage. The mothers who decided to stay in Bayou Oaks and anticipated possible future flooding took multiple steps to increase their family's safety before the next flood. Meghan said: "So, I want to get four life jackets, I want to get two boats, and—you know, because you never know. Like what if we have to get out of here, right? There's no—you might be safe here, or you might not be safe here, right?" Sarah also planned to stock up so she would be ready for the next flood: "Because we don't have life jackets. I mean all the things that I'm going to get. Like we will have a kayak, which is ridiculous, but we will have a kayak. We will have life vests for our children."

Several mothers also took steps to make future flooding easier the next time, like Ruth, who made conscious decisions about her garage, the only part of her new, elevated home that would remain at risk:

> Like the walls of the garage. You know people do drywall there. I'm like, "We're not doing drywall there. I'm not ripping shit out once a year—twice a year. I will plan for a flood." That's fine. I'm not going to redo anything. So we're doing concrete blocks and sealing the concrete blocks so that nothing has to change there. I mean I'm just not. I'm not going to play the game. 'Cause it's a game. Like that's what it is. The city is like fucking with us. You know like the city's got their stuff that they have to deal with. I think this is the reality of climate change, and Houston, and all that stuff.

With lifts for their cars, the family could protect their home and vehicles, but the garage walls would still remain vulnerable. So Ruth was determined to make cleanup and recovery as simple as possible the next time it happened. She believed she could not count on the city's mitigation efforts. The mothers also made interior design and furniture decisions based on the probability of future flooding. Out were skirted couches and chairs; in were couches and chairs with metal legs. Out were wood floors, which buckled after a flood; in were wood-look tile floors. Out were cabinets that extended all the way to the floor; in were cabinets that floated up eight inches off the floor. As Rebecca put it, when considering what new furniture to purchase:

This is the thing. People go, "Lawn furniture? Like what would be good." But as good as my taste is, and as much as I like nice things, I . . . I don't feel like I'm gonna own anything long-term. I feel like everything's gotta be a short-term purchase because long-term I'm not gonna have it. So, what's the easiest to take to the street? What costs me the least? What's gonna be the easiest to take apart? Like Ikea, they're gonna have the same stuff, so I can just keep buying the same thing over and over again. I know it'll fit. I just have a shopping list already for the next one.

The Bayou Oaks mothers, ever the logistics managers for their households, made smart choices to ease their recovery just in case their homes were flooded again. These actions, then, more than their words, indicated their expectations for future flooding. Even though most of the first timers said they thought "another Harvey" was unlikely, they still made decisions based on that "unlikely" possibility. The mothers had chosen Bayou Oaks initially as a way to curate their family lives. They saw it as the best place to raise their children in the entire city. Not only did that not change after three floods in three years, for most of the mothers, but they justified the decision to stay as being on behalf of their children. Although a few mothers fretted about the potential impact on their children of experiencing yet another flood, on balance they believed the risk was worth it for the chance to continue living in the perfect neighborhood.

Conclusion

The brand new, seven-foot-elevated Bayou Oaks Elementary opened, finally, on October 19, 2020, just over three years after Hurricane Harvey sent three feet of water through the old building. Its planned August 2020 opening had been delayed not by another flood, but by the COVID-19 pandemic. The new two-story, tan brick building looked nothing like the original, charming but ramshackle school; it looked like something one might find out in the well-resourced suburbs. The two-story structure was massive, much bigger than the old building, and beautiful gardens were still under construction out back. After the first six virtual weeks of the school year, the Houston school district gave parents the option of having their children continue to learn virtually or in person. Across the district, only 54 percent of elementary school children returned to school in person, but fully 73 percent of Bayou Oaks elementary children went back in person, ranking the school second out of all 160 elementary schools in the Houston district. There is no doubt that the excitement over the new building and the mothers' longing for them to be back were a big part of why the families were eager for their children to return to in-person school. They were hungry for normalcy, and the brand new building was a balm for the neighborhood. As Leah, mother of two, explained: "ALL the feelings today as we walk to school for the first time [in over three years]. Getting back to our school always felt like the final piece of putting our lives back together after Harvey."

While by fall of 2020 nearly all the mothers were back into renovated or rebuilt homes, nothing felt final until the school reopened. To understand why Leah and many other mothers felt this way, the school must be viewed as a core component of family life, a key neighborhood institution that threaded itself through the families in ways that increased social cohesion as well as neighborhood attachment. The school's reopening, bigger and better than ever, felt like confirmation that the mothers had made the right choice to adapt and to stay in Bayou Oaks. On the first day back, masked children streamed into the building, welcomed by waving teachers and administrators. For the Bayou Oaks mothers, this was what they had waited for so long. For three years they'd had to drive two miles north to a subpar facility; now, they could walk their children to school as they could before Harvey. That's not to say that there wasn't anxiety about the possibility of contracting COVID. Most of the mothers who sent their children back were concerned, but they decided that the benefits outweighed the risks. And it wasn't just going back to school after a longer summer than usual, as children all over the district were doing. This was part of "putting [their] lives back together." The school was back, signaling that so was the Bayou Oaks community.

Although there were a few close calls for home flooding in Bayou Oaks after Harvey, and anxiety was high throughout the record-breaking 2020 Gulf of Mexico hurricane season, the neighborhood survived the following three years without a flood. As the globe warms, high-intensity storms and flooding will become more frequent events, and not just in low-lying areas like Houston and Bayou

Oaks. Recent estimates suggest that two to three times more Americans are exposed to serious flooding than previously realized, and that more than forty million Americans currently live in high-risk areas. Moreover, due to population growth patterns and increased economic activity, the number of Americans at risk will only continue to grow.[1] Flood mitigation efforts and adaptive strategies are vital, and humans will spend more time combating water in the coming decades. According to David Wallace-Wells, just two degrees Celsius of warming will result in 50 percent more flooding deaths every year.[2] The question of what to do with homes and neighborhoods in harm's way is a very difficult one, and families and how they operate seem absent from much of the public debate.

If our best strategy to cope with the loss of neighborhoods is "managed retreat," or systematic policy efforts to move people out of flood-prone areas, this book shows the potential challenges to this approach, especially among families with school-aged children.[3] Policies that ignore the reasons families select neighborhoods, or that don't take into account how neighborhoods can be core parts of identities, will not succeed. Too little of the public discourse around moving people from flood zones takes into account how families actually make decisions about where to live and whether to leave. Cold economic calculations are one thing; a mother's desire for her child to attend a particular school in a complicated school system is quite another. As the book also shows, families are not making these decisions in a vacuum; they are listening to and being influenced by their neighbors. Communal decision-making about retreating from flood-prone areas has a strong influence on individual decisions.[4] If most people are staying, residents feel the pull to stay; if most are leaving, they feel the desire to leave. Policies that support the relocation of entire communities may be better embraced by residents, though this book shows the importance of also considering the life stage of families as well as school zones.[5] As Jeff Goodell notes, "the biggest issue any relocation strategy will have to overcome is simply that people love their homes and don't want to leave."[6]

That love for their home and neighborhood was strong in the Bayou Oaks community, and twenty-eight out of thirty-six mothers decided to stay after Harvey. Just over half of those who stayed made structural home adaptations anticipating future flooding, either elevating their existing homes or rebuilding something higher. The rest, mostly first-time flooders, decided to just remodel and bet that they would not be flooded again. The number of times families had experienced home flooding was strongly related to their decision about what to do with their homes. While sixteen mothers chose to just remodel their homes, staying on ground level, they were largely first-time flooders (see table 1). Twelve mothers chose to adapt to the neighborhood's flooding potential by lifting up their existing homes or tearing them down and rebuilding something higher. The remaining eight mothers chose to move, six of them to nearby neighborhoods. Of the five mothers who had been flooded three times, none remodeled; they all either adapted by lifting up or rebuilding, or they moved. With only one exception, the movers no longer had children attending Bayou Oaks Elementary, so their ties to the neighborhood may have been less strong than those of the other families.

That so few mothers decided to move out of Bayou Oaks, including eleven families who had been flooded at least twice in three years, is initially surprising. Many of the mothers who were flooded for the first time during Harvey believed that flood to be a fluke, despite the prior two floods that had impacted other parts of the neighborhood. This is especially interesting because virtually all of the mothers understood and endorsed the concept of climate change, including how it could impact Bayou Oaks. The mothers in this book, then, were not climate change deniers, despite their decision to stay in the neighborhood, and despite many of the families having links to the oil and gas industry. The mothers "believed in" climate change, and also that their neighborhood's troubles directly resulted from it. In fact, they were pessimistic about the future of not just their neighborhood but also their city as a whole. Deborah, one

Table 1 Bayou Oaks Mothers' Housing Decisions after Hurricane Harvey

Number of Times Flooded	Moved	Lifted/Rebuilt	Remodeled
1	2	3	14
2	4	6	2
3	2	3	0

of the mothers who moved out of Bayou Oaks and into an elevated home nearby, said: "I think climate change is going to subsume the entire city of Houston." Or as Allison, a twice-flooded mother, put it: "We worry about climate change and what the future of Houston is economically. I'm not a climate denier person. I'm a science person ... even though [my husband] works in oil and gas. I mean, we are aware of these issues. It's just I ... I believe in the predictions." Rebecca recalled a conversation with her teenaged son shortly after their second flood, in which he asked if the neighborhood was going to keep flooding:

> I said maybe. I said maybe. I said, "Honey, you know climate change is real. The good news is that just because this year's been really tough for hurricanes doesn't mean next year will be." He goes, "Could it get worse?" And I said "It could. It could get worse. It could get better. We don't know. We have to ... we have to be ... we have to adapt. And we just have to take what we're given because these aren't things we can control."

One three-time flooded mother emphatically put it this way:

> I feel anxious about the future of Houston and every coastal community in the country. Climate change is real and if I hear one more Republican economist tell me about science, it's like, "OK, you know nothing." You know, Trump saying, "I think the weather's going to change back." Don't talk to ME about science!!!

So while the mothers not only believed climate change was real, and that their city and neighborhood were likely to keep flooding, these beliefs did not, by and large, shake them from their determination to stay. This paradox—simultaneously believing their neighborhood was "doomed" but deciding to stay nonetheless—is one of the puzzles this book tries to understand. As the book reveals, the vast majority stayed because they believed the neighborhood was the best place to raise their children in Houston, and because living in Bayou Oaks was an important part of their identities as urban, progressive mothers. The neighborhood met a series of precise criteria the mothers had about where and how to raise their children, and the local elementary school was a major draw. Bayou Oaks allowed the mothers to live in a high socioeconomic status, mostly White neighborhood but send their children to a racially diverse, public magnet school for foreign languages. The school met their desire to raise culturally sensitive, multilingual children who would be prepared to succeed in a diverse future world. They valued this, and the family-friendly nature of the neighborhood, because these characteristics were part of their motherhood ideals. The choice of Bayou Oaks was thus part of their curation of family life, which they saw as their responsibility and took very seriously. This desire to curate their children's lives so carefully is, of course, at direct odds with the concept of knowingly putting your children at risk of experiencing another significant flooding event. The mothers tended to downplay the risk of future flooding, perhaps as a way of explaining this tension. The risk of another flood, they felt, was less than moving their families out of the neighborhood they had so carefully selected. The desire to curate family life was strong, and the book shows how the mothers continued to try to do so during the storm and in the flood's aftermath.

The concept of curation is central to understanding why the mothers wanted to stay in Bayou Oaks. Related to, but distinct from, intensive mothering and concerted cultivation, curation, I argue, is a set of practices embedded within, connected to, and reinforced by multiple institutional environments, undertaken by mothers to situate their children for future success.[7] Curation entails not just family-level processes, but also how families interact with neighborhoods and schools. To ensure their children's success, mothers carefully selected the neighborhood and school, with an eye toward curating their children's daily lives. This involved not just engaging in the right extracurriculars, but also considerations of peers, both their own and their children's; a school's curriculum; a neighborhood's physical infrastructure; its social connections; and its location vis-à-vis important cultural institutions. They wanted to situate their children in the best possible environment, not only in terms of social class but also the right kind of exposure to diverse others. Mothers continually anticipated needs, planned extensively, laid groundwork for success, and kept their children's well-being and success at the front of their minds. Importantly, they sought not to completely eliminate strife and stress for their children, because they believed challenges were beneficial to child development, but rather to *situate* them in the type of environment that would allow them to find success. They therefore endeavored to curate an environment in which their children would grow and develop. Fathers were not absent from this process, and many were quite involved in their family's day-to-day lives. But the mothers were the logistics officers, the delegators of tasks, the guardians of stability, and the curators.

When their homes were threatened by another flood event, the mothers transferred these curation skills to the flood experience. They individualized their responses to the communal disaster, believing they could single-handedly prepare, protect, and then dig the family back out, and that it was their responsibility to do so. The flood would be just another life event their children had to navigate, and they sought to minimize potential trauma and maximize potential benefits. As Harvey approached, most of the mothers were able to draw upon flood capital, or valuable information about how to prepare for, and mitigate after, a flood, that circulated widely in

Bayou Oaks. While the majority of first-time flooders did not carry out significant preparations because they believed they were unlikely to be flooded, all of the rest of the mothers undertook substantial, and in some cases expensive, preparations. Due to the neighborhood's extensive experience with flooding, the mothers who made preparations knew what to do: elevate furniture and precious items, take everything out of lower cabinets and drawers, roll up rugs, unplug and elevate electronics, and clear the drains in the street, just to name a few tasks. Some mothers went further, going so far as to pack up much of their homes, to have pianos placed in storage, and to preemptively find temporary rental housing. They did these tasks to try to protect their homes and families, but also to make the aftermath easier. These women were planners—logistical experts—and they soaked up this flood capital from their neighbors or distributed it to others.

Most mothers also did this work almost entirely on their own, partly because their husbands were busy with work but also because many of the men did not believe Bayou Oaks was going to flood again. Although in most cases the husbands were home when the storm hit, the mothers still continued to make the contingency plans and evacuation decisions. They were the ones who had plotted out a place to go if they had to leave their homes, and they were the ones who had the power to decide whether to evacuate to the convention center or not. They also felt the chief responsibility for ensuring their children were all right during and after the storm. The mothers downplayed danger, suggested a sense of fun and adventure during evacuations, and neglected their own well-being to tend to their children. Just as during normal times the mothers took responsibility for curating family life, during a disaster they took on the same responsibility. Their evacuation, sheltering, and temporary housing decisions were all driven by what would be best for their families. If they couldn't have what they wanted—their original homes, undamaged by floodwaters—then they would construct the next best thing in a temporary location, while also managing their home's recovery.

Although after the storm a handful of families agonized over whether to return to Bayou Oaks, and a few decided not to, most of the families immediately turned their attention to home recovery. And all of the families quickly started mucking out their homes, with the help of a seemingly endless stream of volunteers. In this phase of recovery, the fathers were engaged in helping to clean out the homes. Most of them took time off work to help, but some could only take a couple of days. Still, the mothers made executive decisions about what to clean and keep and what should be thrown out. They managed large teams of volunteers who converged on Bayou Oaks, including many teachers and staff from Bayou Oaks Elementary. This was when the careful curation of social networks inside and outside the neighborhood and school paid off for the mothers. While teams of volunteers were working in most flooded neighborhoods across Houston, the intersection of class and racial privilege in Bayou Oaks meant that the outpouring of volunteers was actually overwhelming. The families had social connections across many dimensions, including work, their alma maters, and religious institutions, which provided volunteers who could take paid time off from work to help out. In most cases, this flood mitigation work took place before the mothers had had much time to emotionally process what had happened. The mothers pushed through the process on adrenaline and not much else. In addition to their intense stress and the time pressure to clear things out of the house before mold set in, the mothers worried about their children. Many of them chose not to bring their children back to see their destroyed homes, although some felt it was important to show them why they could no longer live there. Regardless of which they decided to do, the reason was the same: to protect the children's emotional well-being.

At the same time, the mothers were also taking steps to maximize their flood insurance payouts, meticulously documenting their losses. Again, the neighborhood's flood capital was valuable, as mothers had detailed information from others on what to do and what not to do. They also tried to be charming, or at least kind, to their insurance adjusters, who were very busy but crucial for the mothers' future financial payouts. The mothers politely persisted if they didn't agree with their settlement offers, some doing hours of research to submit the strongest claims possible. Their skills in navigating bureaucracies—ingratiating themselves with people in power, providing careful documentation, paying attention to detail, and knowing how to politely insist when they were not satisfied—paid off in this situation, as mothers who persisted the most netted tens of thousands of dollars beyond their initial settlements.

In addition to navigating the insurance claims process, the mothers were also navigating the complex realm of gifts or loans from friends and family. In the immediate aftermath of the storm, the mothers' friends and family from all over the world kept up with what was happening via Facebook and the catastrophic images broadcast in the media. They wanted to help their flooded friends and family members. Some reached out privately and offered money or household items, some sent gift cards for Target or Kroger in the mail, and others created GoFundMe sites to raise money for the families. The more public nature of GoFundMe created a divide within the neighborhood about the propriety of accepting such funds. Many mothers felt that since they all had flood insurance, they shouldn't ask for additional funds. Others felt that their friends and family wanted to help, and after all, they could use the cash. These judgments and feelings would last, and feelings about the matter were running high even a year after the storm. Private gifts from family were generally not disclosed to others, and so in some ways were easier to accept, but were no less fraught. Mothers worried about the obligations they would have to those family members going forward, about whether their families could afford the gifts, and about whether they would be expected to repay the gifts. Although most of the mothers did not receive large financial gifts or loans, all of them had access to capital through their social networks of friends and family and could have used it if needed. In addition to these potential financial resources, virtually all of the mothers received substantial in-kind gifts, such as free places to stay, volunteer labor, furniture and household goods, meals dropped off, cars to borrow, and childcare. By virtue of their social class, the mothers' friends and families had resources they could draw upon when needed, and these resources sped the recovery process for many families. This book reveals these webs of support, both financial and social, as well as the mothers' strong bureaucratic navigation skills, as some of the main mechanisms for increasing inequality in a community after a disaster, rather than acting as equalizers.

The mothers' high standards for parenting were already challenging to meet during normal times, much less after a disaster. But they tried to uphold those standards while living in temporary spaces. Some of the mothers prepared family meals on camp stoves or in microwaves, insisting that everyone sit around a table, albeit a tiny, temporary one, for dinner, as they usually did. Many tried to enforce their usual rules limiting their children's screen time, despite their own myriad stresses and obligations. And most tried to keep up with their children's usual extracurricular activities. When they ultimately could not meet the high standards they had for their parenting, they blamed only themselves, not the disaster they had gone through.

In the midst of trying, and failing, to meet their high parentings standards—and for two-thirds of the mothers, continuing to work full time—they also were managing "another full-time job," home recovery. The homes had to be visited daily to pick up mail and packages, to check on the contractors' progress, and to make decisions about fixtures and finishes. Furniture had to be selected and ordered. The insurance adjuster had to be called, documents faxed, and tile selected. With only a couple of exceptions, the mothers were managing this process on their own. At first the mothers viewed this gendered division of labor as inevitable, because they saw the restoration of their homes as within their domain, part of their responsibility as good mothers. Some also viewed their husbands as not capable of managing the process, due to either their work schedules or their ineptitude in managing details. As time passed, however, the mothers began to chafe at what they increasingly viewed as an unfair division of labor. Tensions rose, and marital conflict increased. In typical heterosexual couples with two working parents, women report spending more time on housework but also tend to view their unequal division of labor as reasonable.[8] While this was largely true for the Bayou Oaks mothers as well, adding a ruined home to the mix tipped the balance into a marital conflict zone.

The conflict was so extreme that a handful of the mothers contemplated divorce, and one couple did divorce in the year following Harvey. More typical, though, were frequent arguments over daily hassles, details of the home recovery, and finances. For most of the mothers, these were not things they usually argued about with their husbands, but the post-Harvey experience escalated these tensions. In addition, their temporary homes lacked the privacy they were used to, which increased stress and decreased intimacy. The mothers believed initially that once they were settled back in their renovated or rebuilt homes, their marriages would go back to normal. But for many families, Harvey's effects on marital conflict, like its financial impact, lingered. The mothers' experiences after a disaster illustrated how "sticky" the division of labor is within marriages.[9] The wives, and according to the wives, their husbands, both saw the home as within the domain of the mothers, and thus the restoration of that home also fell into that domain. This division of labor was reinforced by their social networks within the neighborhood and school, where most couples were in the same situation. Both husbands and wives, perhaps, underestimated the time it would take for the mothers to manage that process, especially after the initial mucking out was complete. As the decisions, errands, meetings, and paperwork mounted, the mothers became resentful and angry and took out their frustrations on their husbands. These stressful responsibilities had deep implications for the mothers' lives, and not just for their marital relationships. The mothers experienced numerous mental and physical health effects after Harvey that they attributed to stress. While initially the mothers viewed the home's recovery process as something within their domain of responsibility, as the months passed and the tasks accumulated, they began to feel resentful and stressed, marital conflict increased, and stress took its toll on the mothers' bodies.

In Too Deep tells the story of one group of affluent mothers who lived in one chronically flooded neighborhood in Houston, Texas. While the findings reveal how floods can increase inequality within communities, due to the concentration of resources in higher-income neighborhoods like Bayou Oaks, they also show how difficult it is, even with resources, to recover from losing one's home. The experience takes a toll on the families, especially the mothers. The book also argues that the mothers were so attached to the neighborhood—despite experiencing multiple floods in three years and the likelihood of more flooding in the future—that they assumed enormous costs to stay. It is important to note that this research has limitations. For example, the study focuses on mothers with school-aged children, and the findings cannot be generalized to the entire neighborhood. Although the majority of households in Bayou Oaks are families with children, other types of residents in the neighborhood, especially the elderly, were less likely to decide to stay in Bayou Oaks. In addition, due to the significant experience with flooding the residents had, as well as the knowledge of how to prepare for and how to mitigate after a flood, or flood capital, the findings might be different in a neighborhood that has been flooded for the first time. Finally, in a city or town with a less unequal school system, that is, one with more uniformly excellent schools, families might not be as tied to their neighborhoods after a disaster. Yet while the findings may not be generalizable to the entire neighborhood, to Houston, or indeed to another state, they do illuminate the inner workings of families after a disaster, including how hard mothers work to sustain their high parenting standards even in difficult circumstances. And perhaps most importantly, the book illustrates why families can have strong attachments to a neighborhood, shedding light on the difficult challenges facing us all as climate change accelerates.

When I spoke again with Rebecca in January 2019, seventeen months after Harvey, in an informal follow-up over gin and tonics, she confessed that the entire family was still sleeping on mattresses on the floor. She had not been able to bring herself to get the beds out of storage—she was planning to wait until hurricane season was over—but when we talked, it had been over for several months already. And May and the start of the next hurricane season was now just around the corner. Their storage unit was also still filled with boxes and boxes of household items and keepsakes. "What's the point?" she said, derisively. "It's all just gonna get destroyed if we move it back into the house." Despite this halfway style of living, Rebecca believed that staying was the best thing for her children. She had ensured the future of their schooling by living in this neighborhood, and her family and her children were embedded in the community in ways she judged important. She worried about the future, including the impact on her children of living with the constant threat of flooding, but she also believed that the benefits outweighed the risks. In fact she, like so many of the mothers, wanted to stay *for her children*:

> The reason I don't feel sorry for myself and stuff like that is—not to sound like Mother Teresa over here—but as a parent, I cannot put myself first in this situation. So what I would like to do or something like that really doesn't matter . . . what's best for my son, I think, is for us to stay put. And if I can figure out a way to stay put and yet not flood, that would be amazing but I don't feel like I'm in a [financial] position to do that. So it's like, so which one's worse? To take him out of his community, or to possibly put him through a third flood? At this point he's just like, "This is just life," you know—I mean kids are . . . they can adapt.

Many of the mothers agreed with Rebecca about their children being adaptable and that the threat of future flooding was just something the families had to live with in order to keep their community and schools. They didn't believe there was another place in Houston with the same combination of neighborhood amenities, a strong community, and good schools. Everything the mothers did was for their children; their children's well-being was why they had chosen Bayou Oaks in the first place. It fulfilled their ideal of an urban community with suburban benefits, tree-lined streets, and comfortable three-bedroom homes, still relatively close to downtown. They explicitly rejected the suburbs. The Bayou Oaks location allowed their families to have short commutes and to take advantage of the cultural amenities the city had to offer. And on top of that, most importantly, the local elementary school was racially diverse and would instruct their children in foreign languages, two exposures the mothers highly valued. Once they had moved to Bayou Oaks, they discovered a bonus: a community of like-minded professional, mostly college-educated, mostly politically progressive neighbors. With all of those advantages combined, the mothers believed, this was the ideal environment in which to raise their children. They had found the place to plant their roots; once planted, it was difficult to uproot them. In fact, three floods in three years were not enough to do that. So while they all dreaded their children's having to experience another flood, they also made the decision to stay and justified the choice as being on their children's behalf. They had a narrative frame for why they had selected Bayou Oaks, and they employed that frame to explain the decision to stay.

The story of Bayou Oaks provides a cautionary tale for policy makers. When families believe they have found the best place to live and the best place to raise their children, and especially when good schools are at stake, it will not be easy to convince them to move. I explicitly asked the mothers what they would do if they received a federal buyout payment they thought was fair. They told me they would try to buy a home that was elevated *in Bayou Oaks*. They wanted to stay, especially if they could do so and not have to worry about flooding. Of course, even in an elevated house, the garage is susceptible to flooding. Neighbors on the ground level will be flooded. And in any given driving rainstorm, the streets will flood and become impassable. But those are considered minor problems compared to having to uproot the family. As more and more communities begin to confront climate change, these types of decisions will become more and more frequent. Middle-class and affluent families, especially those with school-aged children, are likely to be rooted in place. These families have the means to choose their homes and communities with care. And as the Bayou Oaks mothers demonstrate, these decisions are bound up in mothering itself. Future responses to climate change cannot be blind to how gender and family are intricately intertwined with decisions about where to live and how to raise children.

Methodological Appendix

I felt compelled to do this research. It's hard to explain it any other way. Just after Harvey, in late August 2017, I drove my children back to Houston from central Texas, where we had evacuated ahead of the storm, leaving my husband at home to deal with potential flooding. I couldn't believe that Bayou Oaks had flooded *again*. We didn't live in Bayou Oaks, but we lived in the next neighborhood to the west, which was much less expensive but, ironically, on higher ground. Our children had attended Bayou Oaks Elementary several years earlier through the luck of Houston's school lottery system, so some of the repeatedly flooded residents in Bayou Oaks were our school acquaintances. My head was full of questions. What would the families do? What would happen to the school community, with the school building now ruined? Were the children all right? What had they experienced during the flood? How quickly could the families get their homes cleared out, and what would they do next? Surely they wouldn't stay in the neighborhood now, after three floods in three years? What would that mean for the future of the neighborhood, and of the school?

Although Harvey's floodwaters lapped at our foundation for three straight days, we were spared the indignity of its invasion into our home and could look ahead to community recovery. Everything was shut down for nearly eight days in Houston after Harvey, leaving time for flood recovery work. While my husband and teenagers mucked out homes and I hosted the Kimbro childcare center and taqueria for flooded Bayou Oaks toddlers, I started to plan my research design and interview guide. I felt that the world needed to know the story of Bayou Oaks, because it was on the front lines of climate change, and because I had witnessed what the flood experience did to a family. Less than two weeks later, thanks to an expedited institutional review board (IRB) process at Rice University for Harvey projects, I was ready to start interviews.

This study examines how three dozen mothers of school-aged children adapted to repetitive flooding in one neighborhood in Houston, Texas. Based on more than two years of research in Bayou Oaks, the project follows the mothers through their flood recovery process and examines risk perceptions; mothering before, during, and after the storm; the long recovery year; the consequences of experiencing flood trauma for the mothers' mental and physical health and relationships and their children's well-being; and mothers' assessments of the future of the neighborhood. In total, I conducted more than one hundred hours of in-depth, semistructured interviews with thirty-six mothers of school-aged children, speaking to them first just after Hurricane Harvey in September and October 2017, and again a year later in the fall of 2018. During the intervening year, I kept in touch with the mothers via email, phone, and social media. In addition, I conducted participant observation at parties, community meetings, religious events, and school functions. I wanted to understand how the allure of Bayou Oaks could persist after it had been flooded multiple times. I wanted to understand how mothers made sense of what had happened to them and what they believed was likely to happen in the future. I wanted to understand the toll that an experience like this could take on a family and investigate how a community with financial and social resources would deploy them to aid recovery. I also interviewed the principal of the local elementary school twice, once about two months after Harvey and again a year later. The school is such a central part of the Bayou Oaks community, and the mothers mentioned her leadership specifically so often, that it seemed essential to have her perspective, and our conversations yielded many valuable insights.

In-depth, semistructured interviews were appropriate for my research questions because I wanted to understand what the mothers felt and believed about their experiences, as well as how they accounted for their decisions after the storm. I also wanted to collect their storm stories, something that would not be possible with other research methods, such as surveys. The in-depth, in-person interviews also allowed me to understand the mothers' emotions about their experiences before, during, and after the storm, something that interviewing is perhaps uniquely suited for.[1] The semistructured interview allowed me to be sure to cover the same broad set of topics with each mother, while still allowing the interview to flow like a conversation. The method also allowed the mothers to take the interview in any direction they liked, a flexible approach that seemed especially important for disaster survivors.[2] In the first set of interviews, within several weeks of Harvey, the interview guide focused on any prior history of flooding, on their storm stories, on their temporary homes, on their children's experiences and reactions, and on their decisions about their homes. In the follow-up interviews, the interview guide focused on decisions about the home and progress, experiences with restoration/rebuilding, experiences with FEMA and with insurance and mortgage companies, their children's longer-term experiences and reactions, marital conflict, and the mothers' physical and mental health. The first round of interviews was much more about money, and the second round much more about marital strife, than I anticipated. This longitudinal approach is still rare in disaster research, but the needs, desires, feelings, and experiences of the mothers differed radically between the first and second rounds of interviews.[3] Interviewing only at one point in time would have missed key parts of the stories.

RECRUITMENT

I recruited the mothers to my study using a modified snowball sampling method. First, I started with the four mothers I knew before the study through Bayou Oaks Elementary. Then I asked them to recommend additional interviewees they did not think I knew, and who did not know each other, and this helped to diversify the ages of the mothers (and children), as well as the social networks, in the group. To be included in the study, mothers had to live in Bayou Oaks, have been flooded during Hurricane Harvey, and have at least one school-aged child. I also purposely recruited for representation from all three flood groups: three timers, who had been flooded during all three neighborhood flood events, Memorial Day 2015, Tax Day 2016, and Harvey; two timers, most of whom had been flooded by the Memorial Day flood and Harvey; and those whose first flood was Harvey. Ultimately, I asked thirty-nine mothers to participate in the study. I never heard back from two after repeated queries, and one declined. That 92 percent of the respondents I asked to participate agreed to spend an hour and a half or more with me sharing their stories, and that all thirty-six participated in the second round of interviews a year later, is one indication of the generosity of my interviewees. That they freely shared their experiences, fears, concerns, conflicts, and grief is another.

While explaining the research consent form that all mothers in the study signed, I also informed the mothers that I would take every precaution to disguise their identities and maintain their anonymity in the book. Bayou Oaks is a pseudonym for the neighborhood (and the elementary school). Almost all of the mothers told me that after the flood, and in particular after strangers, friends, and family had cleaned out their homes, they were an "open book," not concerned about anyone knowing who they were. "We have no privacy anymore," they would say. Nevertheless, I felt a special obligation to maintain their anonymity, above and beyond what my IRB required, even if the mothers were not concerned about it. That all the respondents live in the same neighborhood (of approximately three thousand residents) is an additional consideration, because some of the participants are likely to know each other. Because of this, all mothers and their partners have pseudonyms, and in some cases their number of children, occupations, and other minor details are altered to enhance their anonymity.

INTERVIEWING DISASTER SURVIVORS

Like other disaster researchers, I felt a heighted sense of ethical responsibility to my research subjects, above and beyond the typical rigorous ethical research standards, because they had all recently been through a traumatic experience.[4] As I discovered in the follow-up interviews a year after Harvey, that trauma was still being felt by many of the mothers. Anxiety was high throughout the group, and most of them cried during the first interview in particular (I always had tissues with me). Sometimes I cried with them (more on that in the next section). Although ultimately none of the mothers asked for such help, at each interview, following guidance from my institution's IRB, I had flyers available with information on free trauma counseling and how to get assistance from the Red Cross and FEMA.

My experience interviewing the mothers generally accords with that of other disaster researchers, in that the mothers seemed generally glad to tell their stories, felt like the act of telling them had been useful for them personally, and hoped their participation would be of help to others in the future.[5] Most of the initial interviews, immediately after Harvey, were conducted at my dining room table, the overwhelming preference of my respondents. Although unusual, this was a practical solution to a logistical problem: businesses, libraries, and other public spaces were not open for weeks and even months after the storm, because everything nearby had been flooded. My university office was twenty-five minutes away from Bayou Oaks, with limited parking. And of course at that time the women did not have places of their own to host me, as their own homes were ruined and they were still sorting out temporary housing. Their children were present for a few of these initial interviews but were occupied in other rooms. In one instance, my teenaged son was home and played video games in another room to help occupy the son of a respondent. In another, I met the mother at my office, and while we talked, my teenaged daughter took the mother's children on a walk around campus. In all other instances, my family kindly vacated our home for the duration of the interview.

The second round of interviews, a year after Harvey, was conducted in the women's renovated homes or their temporary rental homes. This allowed me to gain an understanding of their home lives, as well as the latest design trends, since the renovated homes all had brand-new kitchens, most of which were very similar to one another (white subway or hexagonal tiles, quartz countertops, brass finishes, cabinets in shades of blue and gray). Most importantly, it allowed me to link these informal observations of their private spaces with the rich content of their interviews.[6] As the follow-up interviews progressed, it very quickly became clear just how much of the story I would have missed by only interviewing the mothers once. Their circumstances were often radically different a year later, in terms of both their residential situations and their marriages and financial outlooks. Some initially seemed quite settled back into their homes and family lives, though cracks in the façade would generally appear as the interviews progressed. Others were up front about how frazzled they still felt and how worried they were about the future of the neighborhood.

My own experience with the interviews is worth a mention. Like others who have interviewed respondents who have faced disaster, violence, or other traumas, I found it exceedingly difficult to carry around in my head thirty-six traumatic stories.[7] In the mothers' social group, flood stories were unremarkable, and everyone knew someone who'd had it worse. So for them, I was often the only person who had heard their stories from start to finish. It was almost as though the mothers had partially unburdened themselves and put their experiences onto my shoulders. I was more than willing, even eager, to do this if it helped them. I could see that talking to me was at least a little therapeutic for them, and several explicitly stated that at the end of the interview. But I began to see that it was harmful for me. Especially after the first round of interviews, I had trouble sleeping and constantly "replayed" portions of the mothers' stories in my head.

I thought about the stories nonstop throughout the day. This was difficult not only because I was also teaching and conducting research full time during data collection, in a city that had just been devastated by Harvey, but also because I was coping with survivor's guilt and telling myself my feelings were ridiculous because I had not personally been flooded. It was as though I became so emotionally engaged with the interviewees that I had difficulty disengaging to get back to my regular life.[8] I started taking brisk walks after each interview, without listening to music or a podcast as I normally do, to allow time for processing. I also sometimes called a colleague who had experience with emotionally challenging interviews. These strategies helped. The only other thing that helped me was writing. Once I could commit some of the stories to paper, many of which did not end up fitting in the book, I was able to unburden my mind. Mostly. Even now, almost three years later, I still think of "my" mothers often and feel empathetic toward them. I have now moved farther away from Bayou Oaks, so I don't run into them very often. But they, and their children, will always be in my thoughts.

RESPONDENT CHARACTERISTICS

The mothers of Bayou Oaks are an ideal group to assess how chronic flooding impacts affluent families and their decisions about relocating. The mothers were largely homogeneous in terms of educational background (all but one had a college degree or higher), marital status (all but three were married, and those three were cohabiting with a partner), and race/ethnicity (thirty of thirty-six identified as white). Their ages ranged quite a bit, from thirty-four to fifty-eight, although their children were of similar ages. One-third of the mothers I interviewed were Jewish, and Bayou Oaks is a traditionally Jewish neighborhood. Given the high median cost of homes in the neighborhood (about $450,000), there was more income heterogeneity than I had anticipated: annual household incomes ranged from $75,000 to more than $400,000, with a median of $180,000. There were also major differences across the families in terms of wealth, which played out in their decisions about what they could afford to do with their homes after the flood. While the Bayou Oaks study tells us quite a bit about how affluent families respond to disaster and about how important mothers consider their choice of neighborhood and school to be, it does not allow us a comparison group. The mothers were not all White, but most were, and while incomes ranged, they all lived in expensive homes and had professional occupations or were married to husbands who did. This is certainly a limitation of the work and on generalizing the study's conclusions to other types of communities. But the study also allows us a glimpse into the often shrouded world of disaster response and recovery among the affluent. While I would not seek to generalize my findings to flood victims or disaster victims more generally, the mothers' desire to stay in their community because they saw it as beneficial for their children is a factor I would expect to find in other affluent communities that experience disasters, especially those that may recur, like wildfires or flooding.

DATA ANALYSIS

Immediately after each interview, I jotted down field notes detailing the interview setting, the mother's dress and appearance, her general attitude and emotional state, and any other information I deemed pertinent, such as whether pets, children, contractors, or husbands had interrupted us at any point. I also wrote two to three pages after each initial interview describing each mother's "flood narrative," or her experience during Harvey's flooding, so I could familiarize myself with each story and begin to imagine how the book would flow. These narratives formed the basis for the extended "flood stories" in the book, which introduce or follow specific mothers. One especially interesting feature of my data was the fact that some flood stories intersected. For example, only after I interviewed Shannon, Laura, Ellen, and Emily did I realize that all four were describing the same evacuation and extended stay experience at Shannon's home during the flood. This allowed me to gain a full picture of what those hours in a crowded room were like, from each of their perspectives, and to enhance the narrative arc of the shelter/evacuation section. After completing each follow-up interview, I went back to my notes from the first interview and added to my initial thoughts and experiences. I also wrote a paragraph about any differences I had noticed in the mother's demeanor, emotions, and so forth from the first interview to the second.

After having professional verbatim transcription of the interviews done, I openly coded all seventy-two transcripts for key themes and concepts using AtlasTI. Although informed by grounded theory techniques, I used flexible coding techniques largely as described by Deterding and Waters.[9] In an iterative process, themes were then combined or further refined. Once core themes became clearer, I wrote analytical memos for each one and noted where mothers' accounts converged and diverged. From there, I set out to organize the material in a way that would make sense. Although I knew I wanted the book to proceed with a chronological narrative arc—before the storm, during the storm, after the storm—I also had to consider cohesive chapters that wouldn't touch on too many themes at once. Once I realized that the mothers were justifying their choice to stay in Bayou Oaks by referencing their children's futures, I had the overall arc. From there, I focused in on key concepts, including mothers as "guardians of stability," the limitations of intensive mothering during a disaster, and the consequences of maintaining high parenting standards when times are tough. In addition to the thematic coding, I also classified mothers into groups based on attributes such as how many times they had been flooded and what they decided to do with their houses.[10] This helped me to see patterns, as attitudes, outcomes, and decisions tended to vary, especially based on how many times the mothers had experienced flooding. Finally, I worked to analyze what my participants were saying, to situate that in a broader context, and to gain understanding of their own worldview without imposing my own values or judgments.[11] Overall, the book illustrates how mothers carefully choose where they live and the school their children attend, and the lengths they will go to maintain their family lives even after experiencing repeated disaster.

MY POSITIONALITY: THE INSIDER'S OUTSIDER

I do not live in Bayou Oaks, but my children had previously attended Bayou Oaks Elementary as magnet students, so before the study began I knew four of the mothers I interviewed. I was not, however, a core part of the Bayou Oaks social network; I had only been inside one of the mothers' homes prior to my interviews, for example. This was likely due to the fact that I didn't live in Bayou Oaks proper, that I'm a decided introvert, and that my work prevented me from engaging with the school community as much as I would have liked to. While my children were attending Bayou Oaks Elementary, I was on the tenure track and therefore heavily engaged with my teaching and research. Although I did not know many of the mothers personally before the project began, I was definitely "one of them" as far as my education, race and class background, and social location. I, too, had children who had previously attended Bayou Oaks Elementary (my youngest child had just moved on to middle school when Harvey hit) and had participated in local extracurricular activities. And we shared many other characteristics; my age was about the median age of the group, and I am white, have a professional occupation, and am married with two children. There is no doubt that my personal and status characteristics helped both with gaining access to my respondents and with their comfort in talking to me. I was an insider in many respects, with visible and invisible tools to draw upon, even if I was not a neighborhood resident or someone they knew before the study.[12] They trusted me because I was part of the "Bayou Oaks Elementary family," even if my children had already graduated. These shared characteristics were helpful for access, but probably also meant that I listened to their stories with somewhat less than ideal distance. I certainly felt a kinship with the women, even when I felt critical of some of the things they said. Mostly, I felt sympathy for what they had been through, which despite the level of resources they enjoyed, was incredibly scary and difficult. And even those who did not expect to be flooded again nonetheless lived with the shadow of doubt as they rebuilt their lives. Even in the midst of the chaos the mothers were experiencing after the storm, they wanted to share their stories. They were hopeful that their experiences might help others in the future. They were frank, and funny, and kind. I am indebted to all of them.

Notes

INTRODUCTION

1. Meyer 2017.
2. Hunn, Dempsey, and Zaveri 2018.
3. Bauerlein 2017.
4. National Hurricane Center 2018.
5. Goodell 2018.
6. Fernandez 2019.
7. Goodell 2018.
8. Milly et al. 2002.
9. Hirabayashi et al. 2013.
10. Wallace-Wells 2019.
11. Fothergill and Peek 2004.
12. Angel et al. 2012.
13. Rovai 1994; Fothergill 2004.
14. Howell and Elliott 2019; Raker 2020.
15. Cooper and Pugh 2020; Owens 2016.
16. Knowles, Sasser, and Garrison 2010; Peek 2008.
17. Fothergill and Peek 2015.
18. Peek and Fothergill 2009.

19. Peek, Morrissey, and Marlatt 2011.
20. Peek 2008.
21. Hays 1998.
22. Arendell 2000.
23. Lareau 2011.
24. Delale-O'Connor et al. 2019 (particularly for Black mothers). Verduzco-Baker 2017; Dow 2019 (have more in common).
25. Lareau 2011 ("concerted cultivation" approach); Manning 2019 (shutting other children out).
26. Lewis and Diamond 2015.
27. Cooper 2014.
28. Cooper 2014.
29. Hays 1998; Shirani, Henwood, and Coltart 2012.
30. Milkie and Warner 2014.
31. Villalobos 2014.
32. Shirani, Henwood, and Coltart 2012.
33. Damaske 2013.
34. Ridgeway and Correll 2004; England 2005.
35. Villalobos 2014.
36. Alway, Belgrave, and Smith 1998.

CHAPTER 1. CHOOSING BAYOU OAKS

1. US Census Bureau 2015.
2. Lareau 2011.
3. Davies and Aurini 2008; Calarco 2018.
4. Hays 1998.
5. Hagerman 2020; Rhodes and Warkentien 2017.
6. Bader, Lareau, and Evans 2019.
7. Hagerman 2020, p. 25.
8. Lareau 2014; Krysan and Crowder 2017, pp. 66–84.
9. Lamont and Fournier 1992.
10. Wong 2007.
11. Sherman 2017.
12. Mose 2016.
13. Clements 2004.
14. Putnam 2001.
15. Marsden 2012 (their friends); Bishop 2009 (their own characteristics and political leanings).

16. Harvey et al. 2020.
17. Lankford and Wyckoff 2001; Kimelberg and Billingham 2012.
18. Goldstein and Hastings 2019; Owens 2016.
19. Kimelberg 2014.
20. Weininger 2014; Reay 2007.
21. Lareau, Evans, and Yee 2016 (multiple factors). DeSena and Ansalone 2009; Kimelberg 2014 (maximize their children's outcomes).
22. Holloway 1998.
23. Holme 2002.
24. Dhingra 2019.
25. Sherman 2019.
26. Manning 2019.
27. Houston Independent School District 2016.
28. Billingham and Hunt 2016 (high minority enrollments); Schneider and Buckley 2002 (racial composition).
29. Cucchiara 2013 ("large-enough" group); Posey-Maddox 2014 ("critical mass").
30. Kimelberg 2014; Cucchiara 2013.
31. Cucchiara and Horvat 2014.
32. Kimelberg 2014.
33. Cucchiara 2013.
34. Hagerman 2020, p. 140.
35. Kimelberg 2014.
36. Posey-Maddox 2014; Hagerman 2020.
37. Reay et al. 2007 ("difficult and uncomfortable"). Halley, Eshleman, and Vijaya 2011; Hagerman 2020 (reinforces racial hierarchies).
38. Kimelberg 2014; Posey-Maddox 2014.

CHAPTER 2. STORM PREPARATIONS

1. Daminger 2019.
2. Enarson, Fothergill, and Peek 2007; Enarson 2001; Fothergill 1996.
3. Enarson and Morrow 1997.
4. Peek and Fothergill 2008.
5. Deutsch 2007 ("learned helplessness"); Fothergill 2004 (preparations were unnecessary); Collins 2019 (annoying but typical); Alway, Belgrave, and Smith 1998 (wives' preparatory labor).
6. Tierney 2014.
7. Kahneman et al. 1982.

8. Tierney 2014.
9. Kasperson et al. 2003.
10. Tierney 2014, p. 53.
11. Fothergill and Peek 2015.
12. Fothergill 2004; Enarson, Fothergill, and Peek 2007.
13. Peek and Fothergill 2008; Haney, Elliott, and Fussell 2010.
14. Enarson and Scanlon 1999; Bateman and Edwards 2002.
15. Fothergill and Peek 2015.

CHAPTER 3. DURING THE STORM

1. Bajak and Olsen 2018.
2. Friedrich 2017.
3. Cardinal and Xie 2018; Bajak and Olsen 2018.
4. Solnit 2010; Haney, Elliott, and Fussell 2010.
5. Sherman 2017; Hagerman 2020.

CHAPTER 4. STORM RECOVERY

1. Fothergill 2004; Enarson 2001; Enarson and Chakrabarti 2009; Phillips and Morrow 2008.
2. Enarson 2001.
3. Erickson 2005 (emotional labor); Daminger 2019 (cognitive labor); Hook 2004 (volunteering labor).
4. Doucet 2000.
5. Granovetter 1977.
6. Fothergill 2004, p. 78.
7. Fothergill and Peek 2004.
8. Angel et al. 2012, pp. 57–58; Patterson 1998, p. 152 (networks of wealthier people). Desmond 2012 ("brittle and fleeting" relationships).
9. Fothergill and Peek 2004; Barnshaw and Trainor 2007.
10. Angel et al. 2012.
11. Fritz 1961; Barton 1970.
12. Barton 1970, p. 219; Storr, Haeffele-Balch, and Grube 2016, pp. 88–89.
13. Barton 1970, pp. 238–242.
14. Angel et al. 2012.
15. Fothergill and Peek 2015, pp. 26–27.
16. Goffman 1959.

17. Berk and Berheide 1977.
18. Fothergill 2003.
19. Fothergill 2003.
20. Solnit 2010.
21. Angel et al. 2012; Enarson and Morrow 1997.
22. Enarson and Fordham 2000.

CHAPTER 5. FAMILY IMPACTS

1. Daminger 2019; Erickson 2005; Doucet 2000.
2. Offer and Schneider 2011.
3. Fothergill 2004.
4. Erickson 2005.
5. Morrow and Enarson 1996.
6. Hochschild and Machung 2012 (responsible for home and family life); Doucet 2015; Erickson 2005 (emotional labor of motherhood).
7. Fussell and Lowe 2014.
8. Blair-Loy 2009, p. 67; Collins 2019, pp. 216–217; Doucet 2015.
9. Daminger 2019.
10. Fothergill 2004; Enarson and Scanlon 1999.
11. Alway, Belgrave, and Smith 1998; Fothergill and Peek 2015; Enarson, Fothergill, and Peek 2007.
12. Hochschild and Machung 2012.
13. Peek and Fothergill 2009.
14. Lai et al. 2015; Peek 2008.
15. Gil 1991; Baggerly and Exum 2007.
16. Blaze and Shwalb 2009.
17. Peek, Morrissey, and Marlatt 2011.
18. Fothergill and Peek 2015, p. 218.
19. Terasaka et al. 2015.
20. Peek and Fothergill 2008.
21. Paxson et al. 2012; Rhodes et al. 2010.

CHAPTER 6. TO STAY OR GO

1. Tierney 2014, p. 5.
2. Scott and Lyman 1968.
3. Stone 2008.
4. Damaske 2011.

5. Collete 2018.
6. Erikson 1976; Ritchie 2012.
7. Fritz 1961.
8. Solnit 2010.
9. Norris et al. 2008.
10. Lareau and Goyette 2014.
11. Sherman 2019.
12. Villalobos 2010.
13. Villalobos 2010.

CONCLUSION

1. Wing et al. 2018.
2. Wallace-Wells 2019.
3. Mach et al. 2019.
4. Binder, Baker, and Barile 2015.
5. Kick et al. 2011.
6. Goodell 2018, p. 278.
7. Hays 1998 (intensive mothering); Lareau 2011 (concerted cultivation).
8. Bartley, Blanton, and Gilliard 2005.
9. Erickson 2005.

METHODOLOGICAL APPENDIX

1. Pugh 2013; Lamont and Swidler 2014.
2. Phillips 2014.
3. Phillips 2014. But for a notable exception, see Fothergill and Peek 2015.
4. Tierney 2019; Brzuzy, Ault, and Segal 1997.
5. Levine 2004.
6. Berezin 2014.
7. Dickson-Swift et al. 2007; Brzuzy, Ault, and Segal 1997.
8. Weiss 1995, p. 125.
9. Deterding and Waters 2018, p. 0049124118799377.
10. Deterding and Waters 2018, p. 0049124118799377.
11. Gerson and Damaske 2020.
12. Reyes 2020.

References

Alway, J., L. L. Belgrave, and K. J. Smith. 1998. "Back to Normal: Gender and Disaster." *Symbolic Interaction* 21(2): 175–195.

Angel, R. J., H. Bell, J. Beausoleil, and L. Lein. 2012. *Community Lost: The State, Civil Society, and Displaced Survivors of Hurricane Katrina*. New York: Cambridge University Press.

Arendell, T. 2000. "Conceiving and Investigating Motherhood: The Decade's Scholarship." *Journal of Marriage and Family* 62(4): 1192–1207.

Bader, M. D., A. Lareau, and S. A. Evans. 2019. "Talk on the Playground: The Neighborhood Context of School Choice." *City and Community* 18(2): 483–508.

Baggerly, J., and H. A. Exum. 2007. "Counseling Children after Natural Disasters: Guidance for Family Therapists." *American Journal of Family Therapy* 36(1): 79–93.

Bajak, F., and L. Olsen. 2018. "Hurricane Harvey's Toxic Impact Deeper Than Public Told." *Associated Press News*, March 23.

Barnshaw, J., and J. Trainor. 2007. "Race, Class, and Capital amidst the Hurricane Katrina Diaspora." In *The Sociology of Katrina: Perspectives on a Modern Catastrophe*, ed. D. L. Brunsma, D. Overfelt, and J. S. Picou, 91–105. Lanham, MD: Rowman and Littlefield.

Bartley, S. J., P. W. Blanton, and J. L. Gilliard. 2005. "Husbands and Wives in Dual-Earner Marriages: Decision-Making, Gender Role Attitudes, Division of Household Labor, and Equity." *Marriage and Family Review* 37(4): 69–94.

Barton, A. H. 1970. *Communities in Disaster*. New York: Anchor Books.

Bateman, J. M., and B. Edwards. 2002. "Gender and Evacuation: A Closer Look at Why Women Are More Likely to Evacuate for Hurricanes." *Natural Hazards Review* 3(3): 107–117.

Bauerlein, V. 2017. "As Houston Begins Clean Up, Residents Face Up to Losses." *Wall Street Journal*, August 31.

Berezin, M. 2014. "How Do We Know What We Mean? Epistemological Dilemmas in Cultural Sociology." *Qualitative Sociology* 37(2): 141–151.

Berk, S. F., and C. W. Berheide. 1977. "Going Backstage: Gaining Access to Observe Household Work." *Sociology of Work and Occupations* 4(1): 27–48.

Billingham, C. M., and M. O. Hunt. 2016. "School Racial Composition and Parental Choice: New Evidence on the Preferences of White Parents in the United States." *Sociology of Education* 89(2): 99–117.

Binder, S. B., C. K. Baker, and J. P. Barile. 2015. "Rebuild or Relocate? Resilience and Postdisaster Decision-Making after Hurricane Sandy." *American Journal of Community Psychology* 56(1–2): 180–196.

Bishop, B. 2009. *The Big Sort: Why the Clustering of Like-Minded America Is Tearing Us Apart*. New York: Houghton Mifflin Harcourt.

Blair-Loy, M. 2009. *Competing Devotions: Career and Family among Women Executives*. Cambridge, MA: Harvard University Press.

Blaze, J. T., and D. W. Shwalb. 2009. "Resource Loss and Relocation: A Follow-Up Study of Adolescents Two Years after Hurricane Katrina." *Psychological Trauma: Theory, Research, Practice, and Policy* 1(4): 312–322.

Brzuzy, S., A. Ault, and E. A. Segal. 1997. "Conducting Qualitative Interviews with Women Survivors of Trauma." *Affilia* 12(1): 76–83.

Calarco, J. M. 2018. *Negotiating Opportunities: How the Middle Class Secures Advantages in School*. New York: Oxford University Press.

Cardinal, C., and Y. Xie. 2018. "Implications of Hurricane Harvey on Environmental Public Health in Harris County, Texas." *Journal of Environmental Health* 81(2): 24–32.

Clements, R. 2004. "An Investigation of the Status of Outdoor Play." *Contemporary Issues in Early Childhood* 5(1): 68–80.

Collete, M. 2018. "Flood Games." *Houston Chronicle*, June 28.

Collins, C. 2019. *Making Motherhood Work: How Women Manage Careers and Caregiving.* Princeton, NJ: Princeton University Press.

Cooper, M. 2014. *Cut Adrift: Families in Insecure Times.* Oakland: University of California Press.

Cooper, M., and A. J. Pugh. 2020. "Families across the Income Spectrum: A Decade in Review." *Journal of Marriage and Family* 82(1): 272–299.

Cucchiara, M. 2013. "'Are We Doing Damage?' Choosing an Urban Public School in an Era of Parental Anxiety." *Anthropology and Education Quarterly* 44(1): 75–93.

Cucchiara, M. B., and E. M. Horvat. 2014. "Choosing Selves: The Salience of Parental Identity in the School Choice Process." *Journal of Education Policy* 29(4): 486–509.

Damaske, S. 2011. *For the Family? How Class and Gender Shape Women's Work.* New York: Oxford University Press.

———. 2013. "Work, Family, and Accounts of Mothers' Lives Using Discourse to Navigate Intensive Mothering Ideals." *Sociology Compass* 7(6): 436–444.

Daminger, A. 2019. "The Cognitive Dimension of Household Labor." *American Sociological Review* 84(4): 609–633.

Davies, S., and J. D. Aurini. 2008. "School Choice as Concerted Cultivation: The Case of Canada." In *The Globalisation of School Choice?*, ed. M. Forsey, S. Davies, and G. Walford, 55–72. Oxford: Symposium Books.

Delale-O'Connor, L., J. P. Huguley, A. Parr, and M.-T. Wang. 2019. "Racialized Compensatory Cultivation: Centering Race in Parental Educational Engagement and Enrichment." *American Educational Research Journal*, December 6. https://doi.org/10.3102/0002831219890575.

DeSena, J. N., and G. Ansalone. 2009. "Gentrification, Schooling and Social Inequality." *Educational Research Quarterly* 33(1): 61–76.

Desmond, M. 2012. "Disposable Ties and the Urban Poor." *American Journal of Sociology* 117(5): 1295–1335.

Deterding, N. M., and M. C. Waters. 2018. "Flexible Coding of In-Depth Interviews: A Twenty-First-Century Approach." *Sociological Methods and Research* 50(2): 708–739.

Deutsch, F. M. 2007. "Undoing Gender." *Gender and Society* 21(1): 106–127.

Dhingra, P. 2019. "Achieving More Than Grades: Morality, Race, and Enrichment Education." *American Journal of Cultural Sociology* 7(3): 275–298.

Dickson-Swift, V., E. L. James, S. Kippen, and P. Liamputtong. 2007. "Doing Sensitive Research: What Challenges Do Qualitative Researchers Face?" *Qualitative Research* 7(3): 327–353.

Doucet, A. 2000. "'There's a Huge Gulf between Me as a Male Carer and Women': Gender, Domestic Responsibility, and the Community as an Institutional Arena." *Community, Work and Family* 3(2): 163–184.

———. 2015. "Parental Responsibilities: Dilemmas of Measurement and Gender Equality." *Journal of Marriage and Family* 77(1): 224–242.

Dow, D. M. 2019. *Mothering while Black: Boundaries and Burdens of Middle-Class Parenthood*. Oakland: University of California Press.

Enarson, E. 2001. "What Women Do: Gendered Labor in the Red River Valley Flood." *Global Environmental Change Part B: Environmental Hazards* 3(1): 1–18.

Enarson, E., and P. G. Dhar Chakrabarti, eds. 2009. *Women, Gender and Disaster: Global Issues and Initiatives*. New Delhi: SAGE Publications India.

Enarson, E., and M. Fordham. 2000. "Lines That Divide, Ties That Bind: Race, Class, and Gender in Women's Flood Recovery in the US and UK." *Australian Journal of Emergency Management* 15(4): 43–52.

Enarson, E., A. Fothergill, and L. Peek. 2007. "Gender and Disaster: Foundations and Directions." In *Handbook of Disaster Research*, ed. Havidán Rodriguez, Enrico L. Quarantelli, and Russell R. Dynes, 130–146. New York: Springer.

Enarson, E., and B. H. Morrow. 1997. "A Gendered Perspective." In *Hurricane Andrew: Ethnicity, Gender and the Sociology of Disasters*, ed. W. G. Peacock, B. H. Morrow, and H. Gladwin, 52–74. New York: Routledge.

Enarson, E., and J. Scanlon. 1999. "Gender Patterns in Flood Evacuation: A Case Study in Canada's Red River Valley." *Applied Behavioral Science Review* 7(2): 103–124.

England, P. 2005. "Emerging Theories of Care Work." *Annual Review of Sociology* 31: 381–399.

Erickson, R. J. 2005. "Why Emotion Work Matters: Sex, Gender, and the Division of Household Labor." *Journal of Marriage and Family* 67(2): 337–351.

Erikson, K. 1976. *Everything in Its Path: Destruction of Community in the Buffalo Creek Flood*. New York: Simon and Schuster.

Fernandez, M. 2019. "Two Years after Hurricane Harvey, One Group Says It Has Been Overlooked: Renters." *New York Times*, October 11.

Fothergill, A. 1996. "Gender, Risk, and Disaster." *International Journal of Mass Emergencies and Disasters* 14(1): 33–56.

———. 2003. "The Stigma of Charity: Gender, Class, and Disaster Assistance." *Sociological Quarterly* 44(4): 659–680.

———. 2004. *Heads above Water: Gender, Class, and Family in the Grand Forks Flood*. Albany: State University of New York Press.

Fothergill, A., and L. A. Peek. 2004. "Poverty and Disasters in the United States: A Review of Recent Sociological Findings." *Natural Hazards* 32(1): 89–110.

———. 2015. *Children of Katrina*. Austin: University of Texas Press.

Friedrich, M. J. 2017. "Determining Health Effects of Hazardous Materials Released during Hurricane Harvey." *Journal of the American Medical Association* 318(23): 2283–2285.

Fritz, C. E. 1961. "Disaster." In *Contemporary Social Problems*, ed. R. K. Merton and R. A. Nisbet, 651–694. New York: Harcourt, Brace and World.

Fussell, E., and S. R. Lowe. 2014. "The Impact of Housing Displacement on the Mental Health of Low-Income Parents after Hurricane Katrina." *Social Science and Medicine* 113: 137–144.

Gerson, K., and S. Damaske. 2020. *The Science and Art of Interviewing*. New York: Oxford University Press.

Gil, E. 1991. *The Healing Power of Play: Working with Abused Children*. New York: Guilford Press.

Goffman, E. 1959. *The Presentation of Self in Everyday Life*. New York: Doubleday.

Goldstein, A., and O. P. Hastings. 2019. "Buying In: Positional Competition, Schools, Income Inequality, and Housing Consumption." *Sociological Science* 6: 416–445.

Goodell, J. 2018. *The Water Will Come: Rising Seas, Sinking Cities, and the Remaking of the Civilized World*. New York: Little, Brown.

Granovetter, M. S. 1977. "The Strength of Weak Ties." In *Social Networks: A Developing Paradigm*, ed. S. Leinhardt, 347–367. New York: Elsevier.

Hagerman, M. A. 2020. *White Kids: Growing Up with Privilege in a Racially Divided America*. New York: NYU Press.

Halley, J., A. Eshleman, and R. M. Vijaya. 2011. *Seeing White: An Introduction to White Privilege and Race*. Lanham, MD: Rowman and Littlefield.

Haney, T. J., J. R. Elliott, and E. Fussell. 2010. "Families and Hurricane Response: Risk, Roles, Resources, Race and Religion." In *The

Sociology of Katrina: Perspectives on a Modern Catastrophe, ed. D. L. Brunsma, D. Overfelt, and J. S. Picou, 77–102. Lanham, MD: Rowman and Littlefield.

Harvey, H., K. Fong, K. Edin, and S. DeLuca. 2020. "Forever Homes and Temporary Stops: Housing Search Logics and Residential Selection." *Social Forces* 98(4): 1498–1523.

Hays, S. 1998. *The Cultural Contradictions of Motherhood*. New Haven, CT: Yale University Press.

Hirabayashi, Y., R. Mahendran, S. Koirala, L. Konoshima, D. Yamazaki, S. Watanabe, H. Kim, and S. Kanae. 2013. "Global Flood Risk under Climate Change." *Nature Climate Change* 3(9): 816–821.

Hochschild, A., and A. Machung. 2012. *The Second Shift: Working Families and the Revolution at Home*. New York: Penguin.

Holloway, S. L. 1998. "Local Childcare Cultures: Moral Geographies of Mothering and the Social Organisation of Pre-school Education." *Gender, Place and Culture: A Journal of Feminist Geography* 5(1): 29–53.

Holme, J. J. 2002. "Buying Homes, Buying Schools: School Choice and the Social Construction of School Quality." *Harvard Educational Review* 72(2): 177–206.

Hook, J. L. 2004. "Reconsidering the Division of Household Labor: Incorporating Volunteer Work and Informal Support." *Journal of Marriage and Family* 66(1): 101–117.

Houston Independent School District. 2016. "Student Profile."

Howell, J., and J. R. Elliott. 2019. "Damages Done: The Longitudinal Impacts of Natural Hazards on Wealth Inequality in the United States." *Social Problems* 66(3): 448–467.

Hunn, D., M. Dempsey, and M. Zaveri. 2018. "Harvey's Floods." *Houston Chronicle*, March 30.

Kahneman, D., S. P. Slovic, P. Slovic, and A. Tversky. 1982. *Judgment under Uncertainty: Heuristics and Biases*. Cambridge: Cambridge University Press.

Kasperson, J. X., R. E. Kasperson, N. Pidgeon, and P. Slovic. 2003. "The Social Amplification of Risk: Assessing Fifteen Years of Research and Theory." In *The Social Amplification of Risk*, ed. Nick Pidgeon, Roger E. Kasperson, and Paul Slovic, 13–46. Cambridge: Cambridge University Press.

Kick, E. L., J. C. Fraser, G. M. Fulkerson, L. A. McKinney, and D. H. De Vries. 2011. "Repetitive Flood Victims and Acceptance of

FEMA Mitigation Offers: An Analysis with Community-System Policy Implications." *Disasters* 35(3): 510–539.

Kimelberg, S. M. 2014. "Middle-Class Parents, Risk, and Urban Public Schools." In *Choosing Homes, Choosing Schools*, ed. A. Lareau and K. Goyette, 207–236. New York: Russell Sage Foundation.

Kimelberg S. M., and C. Billingham. 2012. "Attitudes toward Diversity and the School Choice Process: Middle-Class Parents in a Segregated Urban School District." *Urban Education* 48(2): 198–231.

Knowles, R., D. D. Sasser, and M. B. Garrison. 2010. "Family Resilience and Resiliency Following Hurricane Katrina." In *Helping Families and Communities Recover from Disaster: Lessons Learned from Hurricane Katrina and Its Aftermath*, ed. R. P. Kilmer, V. Gil-Rivas, R. G. Tedeschi, and L. G. Calhoun, 97–115. Washington, DC: American Psychological Association.

Krysan, M., and K. Crowder. 2017. *Cycle of Segregation: Social Processes and Residential Stratification*. New York: Russell Sage Foundation.

Lai, B. S., B. Beaulieu, C. E. Ogokeh, S. Self-Brown, and M. L. Kelley. 2015, August. "Mother and Child Reports of Hurricane Related Stressors: Data from a Sample of Families Exposed to Hurricane Katrina." *Child and Youth Care Forum* 44(4): 549–565.

Lamont, M., and M. Fournier. 1992. *Cultivating Differences: Symbolic Boundaries and the Making of Inequality*. Chicago: University of Chicago Press.

Lamont, M., and A. Swidler. 2014. "Methodological Pluralism and the Possibilities and Limits of Interviewing." *Qualitative Sociology* 37(2): 153–171.

Lankford, H., and J. Wyckoff. 2001. "Who Would Be Left Behind by Enhanced Private School Choice?" *Journal of Urban Economics* 50(2): 288–312.

Lareau, A. 2011. *Unequal Childhoods: Class, Race, and Family Life, with an Update a Decade Later*. Berkeley: University of California Press.

———. 2014. "Schools, Housing, and the Reproduction of Inequality." In *Choosing Homes, Choosing Schools*, ed. A. Lareau and K. Goyette, 169–206. New York: Russell Sage Foundation.

Lareau, A., S. Adia Evans, and A. Yee. 2016. "The Rules of the Game and the Uncertain Transmission of Advantage: Middle-Class Parents' Search for an Urban Kindergarten." *Sociology of Education* 89(4): 279–299.

Lareau, A., and K. Goyette, eds. 2014. *Choosing Homes, Choosing Schools*. New York: Russell Sage Foundation.

Levine, C. 2004. "The Concept of Vulnerability in Disaster Research." *Journal of Traumatic Stress: Official Publication of the International Society for Traumatic Stress Studies* 17(5): 395–402.

Lewis, A. E., and J. B. Diamond. 2015. *Despite the Best Intentions: How Racial Inequality Thrives in Good Schools*. New York: Oxford University Press.

Mach, K. J., C. M. Kraan, M. Hino, A. R. Siders, E. M. Johnston, and C. B. Field. 2019. "Managed Retreat through Voluntary Buyouts of flood-Prone Properties." *Science Advances* 5(10): eaax8995.

Manning, A. 2019. "The Age of Concerted Cultivation: A Racial Analysis of Parental Repertoires and Childhood Activities." *Du Bois Review: Social Science Research on Race* 16(1): 5–35.

Marsden, P. V. 2012. *Social Trends in American Life: Findings from the General Social Survey since 1972*. Princeton, NJ: Princeton University Press.

Meyer, R. 2017. "Hurricane Harvey Is the Wettest Atlantic Hurricane Ever Measured." *The Atlantic*, August 29.

Milkie, M. A., and C. H. Warner. 2014. "Status Safeguarding: Mothers' Work to Secure Children's Place in the Status Hierarchy." In *Intensive Mothering: The Cultural Contradictions of Modern Motherhood*, ed. L. R. Ennis, 66–85. Bradford, ON: Demeter Press.

Milly, P. C. D., R. T. Wetherald, K. A. Dunne, and T. L. Delworth. 2002. "Increasing Risk of Great Floods in a Changing Climate." *Nature* 415(6871): 514–517.

Morrow, B. H., and E. Enarson. 1996. "Hurricane Andrew through Women's Eyes." *International Journal of Mass Emergencies and Disasters* 14(1): 5–22.

Mose, T. R. 2016. *The Playdate: Parents, Children, and the New Expectations of Play*. New York: New York University Press.

National Hurricane Center. 2018. "Costliest U.S. Tropical Cyclones Tables Updated." www.nhc.noaa.gov/news/UpdatedCostliest.pdf.

Norris, F. H., S. P. Stevens, B. Pfefferbaum, K. F. Wyche, and R. L. Pfefferbaum. 2008. "Community Resilience as a Metaphor, Theory, Set of Capacities, and Strategy for Disaster Readiness." *American Journal of Community Psychology* 41(1–2): 127–150.

Offer, S., and B. Schneider. 2011. "Revisiting the Gender Gap in Time-Use Patterns: Multitasking and Well-Being among Mothers and Fathers in Dual-Earner Families." *American Sociological Review* 76(6): 809–833.

Owens, A. 2016. "Inequality in Children's Contexts: Income Segregation of Households with and without Children." *American Sociological Review* 81(3): 549–574.

Patterson, O. 1998. *Rituals of Blood: Consequences of Slavery in Two American Centuries.* New York: Civitas/Counterpoint.

Paxson, C., E. Fussell, J. Rhodes, and M. Waters. 2012. "Five Years Later: Recovery from Post Traumatic Stress and Psychological Distress among Low-Income Mothers Affected by Hurricane Katrina." *Social Science and Medicine* 74(2): 150–157.

Peek, L. 2008. "Children and Disasters: Understanding Vulnerability, Developing Capacities, and Promoting Resilience—An Introduction." *Children Youth and Environments* 18(1): 1–29.

Peek, L., and A. Fothergill. 2008. "Displacement, Gender, and the Challenges of Parenting after Hurricane Katrina." *NWSA Journal* 20(3): 69–105.

———. 2009. "Parenting in the Wake of Disaster: Mothers and Fathers Respond to Hurricane Katrina." In *Women, Gender and Disaster: Global Issues and Initiatives*, ed. E. Enarson and P. G. Dhar Chakrabarti, 112–130. New Delhi: SAGE Publications India.

Peek, L., B. Morrissey, and H. Marlatt. 2011. "Disaster Hits Home: A Model of Displaced Family Adjustment after Hurricane Katrina." *Journal of Family Issues* 32(10): 1371–1396.

Phillips, B. D. 2014. *Qualitative Disaster Research.* New York: Oxford University Press.

Phillips, B. D., and B. H. Morrow. 2008. *Women and Disasters: From Theory to Practice.* Bloomington, IN: Xlibris.

Posey-Maddox, L. 2014. *When Middle-Class Parents Choose Urban Schools: Class, Race, and the Challenge of Equity in Public Education.* Chicago: University of Chicago Press.

Pugh, A. J. 2013. "What Good Are Interviews for Thinking about Culture? Demystifying Interpretive Analysis." *American Journal of Cultural Sociology* 1(1): 42–68.

Putnam, R. D. 2001. *Bowling Alone: The Collapse and Revival of American Community.* New York: Simon and Schuster.

Raker, E. J. 2020. "Natural Hazards, Disasters, and Demographic Change: The Case of Severe Tornadoes in the United States, 1980–2010." *Demography* 57: 653–674.

Reay, D. 2007. "'Unruly Places': Inner-City Comprehensives, Middle-Class Imaginaries and Working-Class Children." *Urban Studies* 44(7): 1191–1201.

Reay, D., S. Hollingworth, K. Williams, G. Crozier, F. Jamieson, D. James, and P. Beedell. 2007. "'A Darker Shade of Pale?' Whiteness, the Middle Classes and Multi-ethnic Inner City Schooling." *Sociology* 41(6): 1041–1060.

Reyes, V. 2020. "Ethnographic Toolkit: Strategic Positionality and Researchers' Visible and Invisible Tools in Field Research." *Ethnography* 21(2): 220–240.

Rhodes, A., and S. Warkentien. 2017. "Unwrapping the Suburban 'Package Deal' Race, Class, and School Access." *American Educational Research Journal* 54(1_suppl): 168S–189S.

Rhodes, J., C. Chan, C. Paxson, C. E. Rouse, M. Waters, and E. Fussell. 2010. "The Impact of Hurricane Katrina on the Mental and Physical Health of Low-Income Parents in New Orleans." *American Journal of Orthopsychiatry* 80(2): 237–247.

Ridgeway, C. L., and S. J. Correll. 2004. "Motherhood as a Status Characteristic." *Journal of Social Issues* 60(4): 683–700.

Ritchie, L. A. 2012. "Individual Stress, Collective Trauma, and Social Capital in the Wake of the *Exxon Valdez* Oil Spill." *Sociological Inquiry* 82(2): 187–211.

Rovai, E. 1994. "The Social Geography of Disaster Recovery: Differential Community Response to the North Coast Earthquakes." *Yearbook of the Association of Pacific Coast Geographers* 56(1): 49–74.

Schneider, M., and J. Buckley. 2002. "What Do Parents Want from Schools? Evidence from the Internet." *Educational Evaluation and Policy Analysis* 24(2): 133–144.

Scott, M. B., and S. M. Lyman. 1968. "Accounts." *American Sociological Review* 33(1): 46–62.

Sherman, R. 2017. "Conflicted Cultivation: Parenting, Privilege, and Moral Worth in Wealthy New York Families." *American Journal of Cultural Sociology* 5(1–2): 1–33.

———. 2019. *Uneasy Street: The Anxieties of Affluence*. Princeton, NJ: Princeton University Press.

Shirani, F., K. Henwood, and C. Coltart. 2012. "Meeting the Challenges of Intensive Parenting Culture: Gender, Risk Management and the Moral Parent." *Sociology* 46(1): 25–40.

Solnit, R. 2010. *A Paradise Built in Hell: The Extraordinary Communities That Arise in Disaster*. New York: Penguin.

Stone, P. 2008. *Opting Out? Why Women Really Quit Careers and Head Home*. Berkeley: University of California Press.

Storr, V. H., S. Haeffele-Balch, and L. E. Grube. 2016. *Community Revival in the Wake of Disaster: Lessons in Local Entrepreneurship*. New York: Springer.

Terasaka, A., Y. Tachibana, M. Okuyama, and T. Igarashi. 2015. "Posttraumatic Stress Disorder in Children Following Natural Disasters: A Systematic Review of the Long-Term Follow-Up Studies." *International Journal of Child, Youth and Family Studies* 6(1): 111–133.

Tierney, K. 2014. *The Social Roots of Risk: Producing Disasters, Promoting Resilience*. Stanford, CA: Stanford University Press.

———. 2019. *Disasters: A Sociological Approach*. Cambridge, UK: Polity Press.

U.S. Census Bureau. 2015. "American Community Survey, 5-Year Estimates." Washington, DC.

Verduzco-Baker, L. 2017. "'I Don't Want Them to Be a Statistic': Mothering Practices of Low-Income Mothers." *Journal of Family Issues* 38(7): 1010–1038.

Villalobos, Ana. 2010. "Mothering in Fear: How Living in an Insecure-Feeling World Affects Parenting." In *Twenty-First Century Motherhood: Experience, Identity, Policy, Agency*, ed. A. O'Reilly, 57–71. New York: Columbia University Press.

———. 2014. *Motherload: Making It All Better in Insecure Times*. Oakland: University of California Press.

Wallace-Wells, D. 2019. *The Uninhabitable Earth: Life after Warming*. New York: Tim Duggan Books.

Weininger, E. B. 2014. "School Choice in an Urban Setting." In *Choosing Homes, Choosing Schools*, ed. A. Lareau and K. Goyette, 268–294. New York: Russell Sage Foundation.

Weiss, R. S. 1995. *Learning from Strangers: The Art and Method of Qualitative Interview Studies*. New York: Simon and Schuster.

Wing, O. E., P. D. Bates, A. M. Smith, C. C. Sampson, K. A. Johnson, J. Fargione, and P. Morefield. 2018. "Estimates of Present and Future Flood Risk in the Conterminous United States." *Environmental Research Letters* 13(3): 034023.

Wong, C. J. 2007. "'Little' and 'Big' Pictures in Our Heads: Race, Local Context, and Innumeracy about Racial Groups in the United States." *Public Opinion Quarterly* 71(3): 392–412.

Index

Allison, 28, 73; belief in climate change, 211; experiences during the storm recovery, 121–22

Amy and Mark, 154, 188; experiences during the storm itself, 78–80, 85; experiences during the storm recovery, 125–26, 128; minimal storm preparations of, 55–56; on the possibility of future flooding, 203; on the possible beneficial lessons for her children from the storm experience, 194–95

Andrea, 25

Anna: impact of the storm on her family, 151–52; reasons for moving from Bayou Oaks after the storm, 202; risk perception of, 53–54

Ashley, 36; on the anxiety of mothers and children on returning to school, 147–49; and the Bayou Oaks North project, 144–47; personal loss felt by because of Hurricane Katrina, 114. *See also* Bayou Oaks Elementary School, recovery/rebuild of

Barton, C. E., 122

Bayou Oaks, 3–4, 11; as a "close in" desirable neighborhood, 18, 20; and the "concerted cultivation" approach to middle-class parenting, 19–20; demographics of (85% non-Hispanic White), 27; diversity in, 18, 25; domination of one-story homes in, 90; home prices in, 18, 19, 26; initial large Jewish population of, 18; origins of, 18, 43; proximity of to downtown Houston, 22, 23; strengthening of social connections in bolstered by Jewish residents and institutions, 31–32; tight-knit nature of the community, 29, 31–32; and the values of urban living, 18–19, 23–24

Bayou Oaks Elementary School, 3, 30, 34; as the core of the Bayou Oaks social infrastructure, 33; as a critical institutional support for the Bayou Oaks community, 115–16; demographics of, 39; enrollment and admissions policies of, 42–43; ethnic diversity of,

251

Bayou Oaks Elementary School (*continued*) 39–42; excellence of, 38; as a magnet school with specialization in diverse languages, 17, 33, 38–39; original construction of, 35–36; Parent-Teacher Organization (PTO) of, 116; passion among the Bayou mothers for their school and "school family," 41–42; in the post-Harvey era, 149–50; as reflecting the personal identities of Bayou Oaks mothers, 37, 40. *See also* Bayou Oaks Elementary School, recovery/rebuild of

Bayou Oaks Elementary School, recovery/rebuild of, 183, 207–8; assessment of damages to the school, 112–13; establishment of a post-Harvey command center, 114–15; teachers' volunteer support for the recovery, 115; volunteer support other than teachers, 116–17

Bayou Oaks mothers, 8–9, 11, 43–44, 217–22; belief in climate change, 210–12; choosing not to send their children to certain schools because they were too White, 39–40; as "city" people rather than "suburb" people, 24–26; comfort and safety the mothers found in Bayou Oaks, 28–29; core values of, 35; disaster preparation by, 14; distinguishing of their neighbors from suburban wealthier neighbors, 27–28; features of Bayou Oaks related to parenting that appealed to the mothers, 20–21; as the "guardians of stability," 5; as household logistics managers, 4, 50, 129, 150, 157–58, 206; importance of Bayou Oaks' location to the mothers, 21–22, 23; justification of the mothers' choices of Bayou Oaks to raise their families, 21; "liberal urbanites" identity of, 40; and the markers of race and class privilege in their neighborhood and school choices, 41; mourning of for their homes before being ravaged by Hurricane Harvey, 75; rejection of other neighborhoods in favor of Bayou Oaks by, 23–24; rejection of private schooling by, 37–38; research and reconnaissance by concerning school choices, 35; and risk mediation for their children, 13; sacrifices made by for their families, 22–23; and social interactions/networking among neighbors, 29–31; and wanting to expose their children to a wider world view, 26–28. *See also* Bayou Oaks Elementary School; Bayou Oaks mothers, experiences of during the storm; intensive mothering ethos

Bayou Oaks mothers, experiences of during the storm: Amy's and Mark's experiences, 78–80, 85; Cathy's experiences, 100, 101; concerns over possible electrocution, 83–84; Deborah's and Peter's experiences, 95–96; discussion among families whether to evacuate their homes and relocate to the George R. Brown Convention Center, 97–102; and the disgusting nature of the flood waters, 81–83; and the disturbing necessity of moving into a friend's or a stranger's two-story home, 91–98; Ellen's and Tony's experiences, 92, 105–6; Emily's and Jim's experiences, 93, 94–96, 97; how the mothers took charge during the storm, 80–81; importance of intensive mothering during the storm, 85–87; importance of managing emotions during the storm, 88–90; Jennifer's experiences, 81–82, 102; Jill and Charlie's experiences, 82; Julia's and Jeremy's experiences, 87, 98–99, 102–5; Laura's and Tony's experiences, 85, 91–92, 97; Leah's experiences, 85–86, 98, 100–101; Mary's experiences, 102; Melissa's and John's experiences, 81, 85–86; Nicole's and Dave's experiences, 83–84; occasional levity during, 84; parenting experiences as the "lead parent" in keeping children safe, 85–86; Rebecca's and Paul's experiences, 81; Shannon's and Sean's experiences as hosts to other families in need, 91–98

Bayou Oaks North, 144–47

Berger, Eric, 53

INDEX   253

Cathy, 115, 162–63; on the decision to stay in Bayou Oaks or move, 198–200; experiences during the storm itself, 100, 101
Children of Katrina (Fothergill and Peek), 10, 173
class, 19, 41, 100, 106; and disasters, 9–11; are affluent communities better off after a disaster?, 10; and households with children, 10–11; role of in social networks, 121–22
climate change, 208, 222; belief of the mothers in, 210–12; flooding as a major aspect of, 7, 209; as a threat to family life, 4–5
"concerted cultivation,"18–19
Cooper, Marianne, 12
COVID-19 pandemic, 207, 208
curated families/community, 4, 13, 21, 70, 75, 76, 81, 85, 90, 109, 153, 193; threats to by catastrophic flooding, 14–15
curation, 19–20; concept of as central to understanding why mothers wanted to stay in Bayou Oaks, 212–14; curation of social networks, 215

Damaske, Sarah, 182
Deborah and Peter, 67, 158; belief in climate change, 210–11; experiences during the storm itself, 95–96; on the possible beneficial lessons for her children from the storm experience, 195–96; reasons for leaving Bayou Oaks after the storm, 202–3
division of labor, 67, 155, 156, 218; conflict and resentment concerning, 162–65, 167, 217–19; efficient, 158–59; gendered, 50, 76, 151, 157, 182

Elaine, on the possible beneficial lessons for her children from the storm experience, 194
Ellen and Bill, 153–54; experiences during the storm itself, 92, 105–6
Emily and Jim: on the decision to rebuild from the ground up after the storm, 186–87; experiences during the storm itself, 93, 94–95, 97; experiences during the storm recovery, 119–20
Enarson, Elaine, 116

Federal Emergency Management Agency (FEMA), 6, 50, 116
flood capital, 50, 111, 213–14, 215–16
flood mitigation efforts, and "managed retreat," 209
Fothergill, Alice, 10, 61, 173
Fritz, A. H., 122

George R. Brown Convention Center (GRB), 96, 101; Julia's and Jeremy's experiences at, 102–5; safety issues concerning, 100
Goodell, Jeff, 209
Granovetter, Mark, 117

Harris County, percentage of covered by a foot or more of water, 6
Houston, 117, 133; flooding in, 5–7; ethnic diversity in, 25; outward expansion of, 18
Houston Independent School District (HISD), 17; hybrid school choice system in, 33; racial and ethnic diversity in, 39; two-tiered system of admissions in, 33–34
Hurricane Harvey (2017), 4, 5–7, 133; as an equal-opportunity flood, 6–7; *Houston Chronicle* investigation in the aftermath of Harvey concerning the NFIP's 50 percent cut-off policy, 184–85; as a massive slowmoving megastorm, 106–7; projected path of, 72; as the wettest Atlantic Hurricane ever recorded, 5
Hurricane Katrina, 9, 10, 97, 122, 178; myths concerning the Superdome as a shelter during, 106; racialized images at the Superdome during, 100
"hurrication"/hurricane parties, 73–74

intensive mothering ethos, 5, 11–13, 19–20, 51, 75, 85–87, 90, 212; among Black mothers, 12; demands of, 155; limitations of, 230; specific components of, 12
intensive parenting ethos, 12–13, 151

Janet, 115, 171
Jennifer, 36, 71, 74–75; indecision over what to do about her ruined home,

Jennifer (*continued*)
178–79; weight gain in the aftermath of the storm, 178
Jessica, 68, 191; impact of the storm on her family, 152–53
Jewish Community Center (JCC), 31, 32, 117
Jill and Charlie: on the community bonds in Bayou Oaks as the reason for not moving after the storm, 190–91; experiences during the storm itself, 82; experiences during the storm recovery, 121; storm preparations of, 63, 69
Julia and Jeremy, 75–76, 158, 178; experiences during the storm itself, 87, 98–99; experiences during the storm recovery, 117–18, 131; experiences staying at the GRB, 102–5; storm preparations of, 45–49, 65, 67

Kelly, 22, 178; on the community bonds in Bayou Oaks as the reason for not moving after the storm, 190

labor, 60, 90, 114, 116, 153, 157; anticipatory labor, 58; cognitive labor, 50, 69, 116, 150, 159; curative labor, 20; devalued labor, 13; emotional labor, 90, 130, 154, 168; gendered labor of family life, 14; organizational labor, 114; physical labor, 51, 70, 116, 124; recovery labor, 14, 154, 160–61, 168, 179; volunteer labor, 116, 127. *See also* division of labor
Lareau, Annette, 19
Laura and Tony, 23–24, 83, 172; experiences during the storm itself, 85, 91–92, 97; experiences during the storm recovery, 141; on the possibility of moving in the future, 191–92; on the possible beneficial lessons for her children from the storm experience, 195; storm preparations of, 54–55, 62
Leah, 30; experiences during the storm itself, 85–86, 98, 100–101; mental stress in the aftermath of the storm, 176
Linder, Jeff, 6

Lucy and Michael: experiences during the storm recovery, 132; storm preparations of, 57–58

Maria, 153; experiences during the storm recovery, 10
Mary, 73, 198; belief public schooling, 37–38; experiences of during the storm itself, 102
Meghan, 159; storm preparations of, 58, 59–60; storm preparations of excluding her husband, 71–72
Melissa and John: depression of Melissa in the aftermath of the storm, 177; experiences during the storm itself, 81; experiences during the storm recovery, 141–42; on the possibility of lifting their house, 185–86; storm preparations of, 56
Memorial Day 2015 flood, 2–3, 52, 54, 107, 171, 174; belief that it was a once-in-a-lifetime storm, 48
Michelle, 23, 32; experiences during the storm recovery, 129, 136
Molly, on the possibility of future flooding, 204
motherhood, 25, 35, 182–83, 212; devalued status of, 13; emotional labor of, 154

Nancy, 175; shock experienced by in the aftermath of the storm, 176–77
National Flood Insurance Program (NFIP), 6; investigation of concerning its 50 percent cut-off policy, 184–85; NFIP Increased Cost of Compliance (ICC) Coverage grants, 185
Nicole and Dave, 16–18, 36, 40; experiences during the storm itself, 83–84; Nicole's inability to sleep in the aftermath of the storm, 178; storm preparations of, 73–74

Paradise Built in Hell, A (Solnit), 139
Peek, Lori, 10, 61, 173

race, 12, 19, 41, 100, 106, 193
racial hierarchies, reinforcement of, 42
racial privilege, 41

INDEX 255

Rebecca and Paul, 1, 44, 220; on the difficult decision whether to stay in Bayou Oaks or move, 180–82; experiences during the storm itself, 81, 107; experiences during the storm recovery, 118–19, 123, 129; and the Memorial Day 2015 flood, 2–4, 107; on the possibility of future flooding, 205–6; storm preparations of, 58

risk: amplification of through "signals" of potential risk, 52–53; influence of social and environmental factors on, 52; "risk climates," 12; risk perception and storm preparations of Bayou Oaks mothers based on prior flooding, 51–54

Ruth, 152–53, 175–76; honesty with her children about the approaching storm, 64–65; meeting of with Bayou Oaks Elementary principal, 36–37; on the possibility of future flooding, 205

Sally, 141
Samantha, 130, 159–60
Sarah, 26, 30, 160; on the community bonds in Bayou Oaks as the reason for not moving after the storm, 190–91; on the possibility of future flooding, 204, 205
"security projects," 12
Shannon and Sean, 172; experiences as hosts to other families in need during the storm itself, 91–98
Sherman, Rachel, 37
sociology, family, 182
Solnit, Rebecca, 139
"status safeguarding," 12
storm impacts on children, 170–76; Cathy's experiences of her children's anxiety and stress, 173–74; children's concern for their mothers, 171–72; children's response to storm impacts through complex language to make sense of the storm, 171; children's response to storm impacts through play, 170–71; decisions by mothers to place a child in therapy, 175; long-term impacts on children, 172–73; Rebecca's experiences concerning her son, 174–75; Ruth's experiences concerning, 175–76

storm impacts on families: anxiety of children and mothers on returning to school after the storm, 147–49; dissatisfaction of the mothers with themselves over the disruption to their children's routines, 153–55; the flood experience as actually valuable to the family children, 193–98; impact of the lack of privacy, 90, 120, 124–25, 156, 165, 218; impact on Anna's family, 151–52; impact on Cathy's family, 168–69; impact on Jessica's family, 152; impact on and the need for mothers to "maintain routines," 151; impact on the well-being of the mothers, 176–79; on restoring normalcy for children and women's division of labor, 151–55. *See also* storm impacts on families' decisions whether to stay in Bayou Oaks or move; storm impacts on married couples' relationships

storm impacts on families' decisions whether to stay in Bayou Oaks or move, 211*tab.*; Anna's decision, 202; Bayou Oaks community and Bayou Oaks Elementary School as the primary reasons for staying put, 189–92; Deborah's and Peter's decision, 202–3; extraordinary measures undertaken by mothers to say in Bayou Oaks, 182–83; four basic choices concerning recovery options for flood-damaged homes (repair the damage and move back in, lift the home, tear down the home and rebuild, sell the property as is and move), 183–87; justification for staying put because "everywhere floods" no matter where one moves, 187–88; number of families that stayed, 210; and the possibility of future flooding, 203–6, 220; Rebecca's and Paul's decision, 180–82; reluctance to leaving the Bayou Oaks neighborhood even among those that did move, 201–2; staying as the only financially responsible choice, 183, storm impacts on families' decisions whether to stay in Bayou Oaks or move (*continued*) 184; strong desire of mothers to stay in the "neighborhood," 192–93

storm impacts on married couples' relationships, 155; Amy's and Mark's experiences, 156; Cathy's experiences, 162–63; conflicts among couples on the division of labor, 156–66; conflicts among couples that threatened the marriage itself, 166–68; Deborah's and Peter's experiences, 158–59, 161–62; impacts on emotional intimacy and sex, 164–65; Julia's and Jeremy's experiences, 158, 163; Meghan's experiences, 159; Rebecca's and Paul's experiences, 164; Samantha's experiences, 159–60; Sarah's experiences, 160; Tara's experiences, 167–68; Tracey's experiences, 155–56

storm preparations before Harvey, 46; assistance of children in storm preparations, 62–63; and the creation of the approaching storm as a "fun adventure" for children, 61, 65–66; emotional stress created by, 51; evacuation of children before the storm, 66; extensive preparations as class-related efficiency, 60; general preparations (logistical, emotional, social) taken on by mothers without the aid of their husbands, 50–51; husbands' reluctance to take part in storm preparations, 67–70; husbands' view of storm preparations, 51; intense/extensive preparations among many Bayou Oaks mothers, 49–50; and the leasing of rental properties outside the flood area, 59; minimal preparations made by first timers, 55–56; preparations of Julia, 45–49; preparations of Laura and Tony, 54–55; preparations of Lucy and Michael, 57–58; preparations of Meghan, 58, 59–60; preparations of Melissa, 56; preparations to protect children from anxiety and fear about the storm, 60–61; preparations of Rebecca and Paul, 58; and risk perceptions based on prior flooding, 51–54; shielding of children from storm preparations, 63–64; spousal conflict over storm preparations and the decision whether to stay in the home or evacuate, 70–72

storm recovery experiences: Allison's experiences, 121–22; Amy's and Mark's experiences, 125–26, 128; and dealing with contractors, 3, 14, 108, 133–34, 140, 157, 158–59, 161, 217; difficulty of the mothers' accepting help in the form of cash, gift cards, and GoFundMe pages, 130–33, 136–38, 216–17; Emily's and Jim's experiences, 119–20, 119–20; experiences of poorer flood victims, 122; and the families' sense of being robbed, 111; financial aid during the recovery, 132–34, 143; financial stresses on the mothers during recovery, 127–30; financial and social resources supporting the recovery process, 109–10; the five-step flood remediation process, 112; fundraising efforts during, 134–35; and the importance of social networks in aiding recovery, 120–22; and the importance of speed in aiding recovery, 118–19; irritation of mothers during, 123–25; Jill's and Charlie's experiences, 121; Julia's and Jeremy's experiences, 117–18, 131; Laura's and Tony's experiences, 141; Lucy's and Michael's experiences, 132; managing role of mothers in, 117–18, 129; Maria's experiences, 120; Meghan's experiences, 111, 118, 141; Melissa's and John's experiences, 141–42; Michelle's experiences, 129, 136; and the problem of scavengers, 126–27; Rebecca's and Paul's experiences, 118–19, 123, 129; return of families to their homes as "gut-wrenching" and "like a war zone," 110–11; the role of flood insurance and FEMA payments in aiding the recovery, 139–42; Tara's experiences, 131. *See also* Bayou Oaks Elementary School, recovery/rebuild of

Tara: experiences during the storm itself, 87–88; experiences during the storm recovery, 131; role of her husband in storm preparations, 69; on the possible beneficial lessons for her children from the storm experience, 193–94; storm preparations of concerning her children, 62–63

Tax Day flood (2016), 48, 52
"therapeutic communities," 122, 189
Tierney, Kathleen, 52

Uneasy Street (Sherman), 37

Wallace-Wells, David, 209
"weak ties," 117, 122–23

Founded in 1893,
UNIVERSITY OF CALIFORNIA PRESS
publishes bold, progressive books and journals
on topics in the arts, humanities, social sciences,
and natural sciences—with a focus on social
justice issues—that inspire thought and action
among readers worldwide.

The UC PRESS FOUNDATION
raises funds to uphold the press's vital role
as an independent, nonprofit publisher, and
receives philanthropic support from a wide
range of individuals and institutions—and from
committed readers like you. To learn more, visit
ucpress.edu/supportus.